T0212422

Lecture Notes in Computer Science   12052

More information about this series at http://www.springer.com/series/7407

Lhassane Idoumghar · Pierrick Legrand ·
Arnaud Liefooghe · Evelyne Lutton ·
Nicolas Monmarché · Marc Schoenauer (Eds.)

Artificial Evolution

14th International Conference, Évolution Artificielle, EA 2019
Mulhouse, France, October 29–30, 2019
Revised Selected Papers

 Springer

Editors
Lhassane Idoumghar
IRIMAS Institute
ENSISA
Mulhouse, France

Pierrick Legrand
Inria Bordeaux Sud-Ouest, IMB
University of Bordeaux
Talence, France

Arnaud Liefooghe
Research Center in Computer Science,
Signal and Automatic Control of Lille
University of Lille
Villeneuve d'Ascq, France

Evelyne Lutton
GMPA
INRA
Thiverval-Grignon, France

Nicolas Monmarché
Laboratoire d'Informatique
University of Tours
Tours, France

Marc Schoenauer
Inria Saclay
University of Paris-Sud
Orsay, France

ISSN 0302-9743   ISSN 1611-3349 (electronic)
Lecture Notes in Computer Science
ISBN 978-3-030-45714-3   ISBN 978-3-030-45715-0 (eBook)
https://doi.org/10.1007/978-3-030-45715-0

LNCS Sublibrary: SL1 – Theoretical Computer Science and General Issues

This Springer imprint is published by the registered company Springer Nature Switzerland AG
The registered company address is: Gewerbestrasse 11, 6330 Cham, Switzerland

Preface

This LNCS volume is made of the best papers presented at the 14th Biennial International Conference on Artificial Evolution (EA[1] 2019), held in Mulhouse, France. This conference proceeds a long series of previous issues, that took place in Paris (2017), Lyon (2015), Bordeaux (2013), Angers (2011), Strasbourg (2009), Tours (2007), Lille (2005), Marseille (2003), Le Creusot (2001), Dunkerque (1999), Nimes (1997), Brest (1995), and Toulouse (1994).

We sought original contributions relevant to Artificial Evolution, including, but not limited to: evolutionary computation, evolutionary optimization, co-evolution, artificial life, population dynamics, theory, algorithmic and modeling, implementations, application of evolutionary paradigms to the real world (industry, biosciences, etc.), other biologically-inspired paradigms (swarm, artificial ants, artificial immune systems, cultural algorithms, etc.), memetic algorithms, multi-objective optimization, constraint handling, parallel algorithms, dynamic optimization, machine learning, and hybridization with other soft computing techniques.

Each submitted paper was reviewed by four members of the International Program Committee. As was the case in previous editions, a selection of the best papers which were presented at the conference and further revised for publication (see LNCS volumes 1063, 1363, 1829, 2310, 2936, 3871, 4926, 5975, 7401, 8752, 9554, and 10764). EA 2019 continued this tradition, selecting high-quality papers for the oral presentation, which amounted in 16 revised papers being included in this volume of the Springer's LNCS series.

As per usual, the EA 2019 success is indebted to dedicated team work, for which I would like to express my gratitude to:

- Edward Keedwell, who accepted to be our keynote speaker
- The Program Committee members for their careful work: the high quality of the selected papers is a proof of their strong commitment
- The Organizing Committee for their efficient work and kind availability, in particular the local team
- The members of the Steering Committee for their valuable assistance
- Aurélien Dumez and Pierrick Legrand for the administration of the conference website
- Anne Jeannin-Girardon, Pierre Parrend, and Marc Schoenauer for their support and management of the MyReview system
- Laetitia Jourdan and Patrick Siarry for publicity
- Pierrick Legrand and Arnaud Liefooghe for editing the proceedings

---

[1] As for previous editions of the conference, the EA acronym is based on the original French name "Évolution Artificielle".

I take this opportunity to thank the different partners whose financial and material support were invaluable

- IRIMAS (Institut de Recherche en Informatique, Mathématiques, Automatique et Signal)
- Faculté des Sciences et Techniques, Université de Haute-Alsace
- Institut Universitaire de Technologies de Mulhouse, Université de Haute-Alsace
- Région Grand-Est
- École Polytechnique de l'Université de Tours
- Inria
- ROADEF
- Association EA

Finally, we are also deeply grateful to all authors who submitted their research work to the conference, and to all attendees who make the conference so lively. The scientific quality as well as the warm and friendly atmosphere of this series of conferences is the result of a rare alchemy that is still maintained. Thank you for all these years of fidelity, thank you for EA 2019.

February 2020   Lhassane Idoumghar

Organization

Chair

Lhassane Idoumghar   University of Haute Alsace, France

Steering Committee

Pierre Collet   University of Strasbourg, France
Pierrick Legrand   University of Bordeaux, France
Evelyne Lutton   INRA Versailles-Grignon, France
Nicolas Monmarché   University of Tours, France
Marc Schoenauer   Inria Saclay, France

Organizing Committee

Bruno Adam   University of Haute Alsace, France
Mathieu Brévilliers   University of Haute Alsace, France
Aurélien Dumez   Inria Bordeaux, France
Germain Forestier   University of Haute Alsace, France
Laetitia Jourdan   University of Lille, France
Anne Jeannin-Girardon   University of Strasbourg, France
Fabrice Lauri   University of Haute Alsace, France
Pierrick Legrand   University of Bordeaux, France
Julien Lepagnot   University of Haute Alsace, France
Arnaud Liefooghe   University of Lille, France
Yvan Maillot   University of Haute Alsace, France
Laurent Moalic   University of Haute Alsace, France
Mahmoud Melkemi   University of Haute Alsace, France
Pierre Parrend   University of Strasbourg, France
Dominique Schmitt   University of Haute Alsace, France
Patrick Siarry   University of Paris-Est Creteil, France
Jonathan Weber   University of Haute Alsace, France

PhD Student Volunteers

Mounir Bendali-Braham
Mokhtar Essaid
Hassan Ismail Fawaz
Soheila Ghambari
Julien Kritter
Hojjat Rakhshani
Imene Zaidi

Program Committee

Hernán Aguirre	Shinshu University, Japan
Christian Blum	Artificial Intelligence Research Institute (IIIA-CSIC), Spain
Stephane Bonnevay	University of Lyon 1, France
Nadia Boukhelifa	INRA, France
Boumaza, Amine	University of Lorraine, France
Nicolas Bredeche	Sorbonne University, France
Mathieu Brevilliers	University of Haute Alsace, France
Nik Noordini Bt Nik Abd Malik	Universiti Teknologi, Malaysia
Stefano Cagnoni	University of Parma, Italy
Francisco Chicano	University of Málaga, Spain
Maurice Clerc	Independent Scholar, France
Manuel Clergue	University of the French West Indies, France
Pierre Collet	University of Strasbourg, France
Fatima Debbat	University of Mascara, Algeria
Laurent Deroussi	University of Clermont-Ferrand, France
Clarisse Dhaenens	University of Lille, France
Carola Doerr	Sorbonne University, France
Marco Dorigo	Université Libre de Bruxelles, Belgium
Marc Ebner	University of Greifswald, Germany
Mounir Elbaz	University of Haute Alsace, France
Rachid Ellaia	Mohamed V-Rabat University, Morocco
Andries Engelbrecht	University of Pretoria, South Africa
Mostafa Ezziyyani	Abdelmalek Essaâdi University, Morocco
Hongying Fei	Shanghai University, China
Francisco Fernandez de la Vega	University of Extremadura, Spain
Cyril Fonlupt	University of the Littoral Opal Coast, France
Germain Forestier	University of Haute Alsace, France
Edgar Galvan	Trinity College of Dublin, UK
Mario Giacobini	University of Turin, Italy
Adrien Goëffon	University of Angers, France
Frédéric Guinand	University of Le Havre, France
Jin-Kao Hao	University of Angers, France
Lhassane Idoumghar	University of Haute Alsace, France
Anne Jeannin-Girardon	University of Strasbourg, France
Laetitia Jourdan	University of Lille, France
Edward Keedwell	University of Exeter, UK
Bill Langdon	University College London, UK
Nurul Mu'azzah Abdul Latiff	Universiti Teknologi, Malaysia

Fabrice Lauri	University of Technology of Belfort-Montbéliard, France
Pierrick Legrand	University of Bordeaux, France
Julien Lepagnot	University of Haute Alsace, France
Jing Liang	Zhengzhou University, China
Arnaud Liefooghe	University of Lille, France
Manuel López-Ibáñez	The University of Manchester, UK
Jean Louchet	Inria Saclay, France
Nuno Lourenço	University of Coimbra, Portugal
Evelyne Lutton	INRA, France
Katherine Malan	University of South Africa, South Africa
Virginie Marion-Poty	University of the Littoral Opal Coast, France
Nicolas Monmarché	University of Tours, France
Una-May O'Reilly	MIT Computer Science and Artificial Intelligence Lab, USA
Gabriela Ochoa	University of Stirling, UK
Damien Olivier	University of Le Havre, France
Luís Paquete	University of Coimbra, Portugal
Andrew Parkes	University of Nottingham, UK
Pierre Parrend	University of Strasbourg, France
Francisco Pereira	University of Coimbra, Portugal
Alain Petrovsky	Télécom Paris, France
Amin Rahati	University of Sistan and Baluchestan, Iran
Frederic Saubion	University of Angers, France
Marc Schoenauer	Inria Saclay, France
Oliver Schütze	CINVESTAV, Mexico
Patrick Siarry	University of Paris-Est Creteil, France
Giovanni Squillero	Politecnico di Torino, Italy
Thomas Stützle	Université Libre de Bruxelles, Belgium
El-ghazali Talbi	University of Lille, France
Eduardo Rodriguez Tello	CINVESTAV, Mexico
Dirk Thierens	Utrecht University, Nederlands
Alberto Tonda	INRA, France
Leonardo Trujillo	Instituto Tecnológico de Tijuana, Mexico
Paulo Urbano	University of Lisboa, Portugal
Sébastien Verel	University of the Littoral Opal Coast, France
Jonathan Weber	University of Haute Alsace, France
Darrell Whitley	Colorado State University, USA
Annie S. Wu	University of Central Florida, USA
Emigdio Z. Flores	Instituto Tecnológico de Tijuana, Mexico
Nicolas Zufferey	University of Geneva, Switzerland

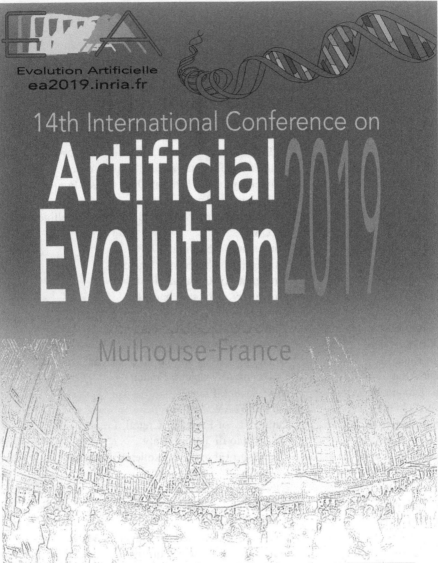

Evolution Artificielle
ea2019.inria.fr

14th International Conference on

Artificial
Evolution 2019

Mulhouse-France

Association
Evolution
Artificielle

 inria
inventeurs du monde numérique

ROADEF

 FST
Faculté des Sciences et Techniques
UNIVERSITÉ HAUTE-ALSACE

Grand Est
ALSACE CHAMPAGNE-ARDENNE LORRAINE
L'Europe s'invente chez nous

IUT Mulhouse
Institut Universitaire de Technologie
UNIVERSITÉ HAUTE-ALSACE

 IRIMAS UNIVERSITÉ
HAUTE-ALSACE
Institut de Recherche en Informatique, Mathématiques, Automatique et Signal

POLYTECH
TOURS
École d'ingénieurs polytechnique
de l'université de Tours

New Directions in Search: Heuristics, Metaheuristics and Hyperheuristics for Real-World Optimisation Problems (Abstract of Invited Talk)

Edward Keedwell

University of Exeter, UK

Abstract. The increasing use of search and optimisation algorithms in real-world applications presents new challenges to researchers to develop algorithms that are computationally efficient and are able to produce meaningful solutions. In this talk, I will describe two approaches that are aiming to address these challenges: interactive evolutionary metaheuristics and sequence-based hyperheuristics. These methods are designed to make use of human intelligence and machine learning to improve search and optimisation performance and to generate feasible solutions for real-world problems in the water industry and operations research problems. Specifically, I will demonstrate an interactive evolutionary algorithm (EA) system that is able to learn human preferences and embed them into the operation of an EA to improve objective and subjective performance criteria. I will then describe recent work in the use of machine learning to understand and create sequences of search operations within a hyperheuristic framework to better understand the problem-algorithm interface and improve search performance.

Contents

From Feature Selection to Continuous Optimization

Hojjat Rakhshani$^{(\boxtimes)}$, Lhassane Idoumghar$^{(\boxtimes)}$, Julien Lepagnot$^{(\boxtimes)}$, and Mathieu Brévilliers$^{(\boxtimes)}$

Université de Haute-Alsace, IRIMAS-UHA, 68093 Mulhouse, France
{hojjat.rakhshani,lhassane.idoumghar,
julien.lepagnot,mathieu.brevilliers}@uha.fr

Abstract. Metaheuristic algorithms (MAs) have seen unprecedented growth thanks to their successful applications in fields including engineering and health sciences. In this work, we investigate the use of a deep learning (DL) model as an alternative tool to do so. The proposed method, called MaNet, is motivated by the fact that most of the DL models often need to solve massive nasty optimization problems consisting of millions of parameters. Feature selection is the main adopted concepts in MaNet that helps the algorithm to skip irrelevant or partially relevant parameters and use those design variables which contribute most to the overall performance. The introduced model is applied on several unimodal and multimodal continuous problems. The experiments indicate that MaNet is able to yield competitive results compared to one of the best hand-designed algorithms for the aforementioned problems, in terms of the solution accuracy and scalability.

Keywords: Metaheuristics · Deep learning · Continuous optimization

1 Introduction

The need for optimization has received a lot of attention in different application areas. Formally, optimization algorithms seek to find a parameter vector x^* so as to minimize a cost function $f(x) : \mathbb{R}^D \to \mathbb{R}$, i.e. $f(x^*) \leq f(x)$ for all $x \in \Omega$, where $\Omega = \mathbb{R}^D$ is the search domain and D is the dimension of the problem. There are no a prior hypothesis about f and optimization algorithms should treat them as black-box functions. This motivated the development of MAs which do not take advantages of problem structure.

MAs are one of the fastest growing fields aimed at solving different complex and highly non linear real-world problems by inspiration from the process of natural evolution or physical processes [9,18]. In MAs, we often have a population of candidate solutions that strive for survival and reproduction. In every iteration, different search operators are applied to the candidate solutions and then the population will be updated based on its success in achieving the goal. Over the last decade, there has been an explosion in the development of a variety

© Springer Nature Switzerland AG 2020
L. Idoumghar et al. (Eds.): EA 2019, LNCS 12052, pp. 1–12, 2020.
https://doi.org/10.1007/978-3-030-45715-0_1

of extensions to further enhance the performance of MAs. However, there are no clear guidelines on the strengths and weaknesses of alternative methods such as the DL models for developing more enhanced optimization algorithms.

The DL approaches use a hierarchy of features in conjunction with several layers to learn complex non-linear mappings between the input and output layer. As opposite to traditional machine learning methods that use handmade features, the important features are discovered automatically and are represented hierarchically. This is known to be the strong point of DL against traditional machine learning approaches. Accordingly, these models have been described as universal learning approaches that are not task specific and can be used to tackle different problems arise in different research domains [1]. In this work, we propose a simple, yet effective approach for numerical optimization based on the DL. The proposed MaNet adopts a Convolutional Neural Network (CNN); which are regularized version of fully-connected neural networks inspired from biological visual systems [12]. The "fully-connectedness" of CNNs enables them to tackle the over-fitting problem and it is reasonable to postulate that they may outperform classical neural networks for difficult optimization tasks.

The rest of the paper is organized as follows. Section 2 presents a review on the related works and describes our motivations. In Sect. 3, we elaborate technical details of the MaNet approach. In Sect. 4, a series of experiments are conducted to show the performance of the introduced method. The last section summarizes the paper and draws conclusions.

2   Related Works and Motivations

The idea of solving optimization problems using neural networks has an old history which has seen a number of advances in recent years [2,3,10,14,23]. In [23], authors developed a Bayesian optimization method, called as DNGO, based on deep neural networks for hyperparameter tuning of large scale problems with expensive evaluation. The main idea is to combine large-scale parallelism with an optimization method to provide an approximate model of the real cost function. They show that DNGO scales in a less dramatic fashion compared to the Gaussian process, while maintains its desirable flexibility and characterization of uncertainty. OptNet [2] is another method proposed for learning optimization tasks by the virtues of DL, sensitivity analysis, bilevel optimization, and implicit differentiation. The authors highlighted the potential power of OptNet networks against existing networks to play mini-Sudoku. In [3], researchers investigated automating the design of an optimization algorithm by Long short-term memory deep networks on a number of tasks. Their results outperform hand-designed competitors for simple convex problems, neural network training and styling images with neural art. Similarly, Li and Malik [14] put forward a deep learning method for automating algorithm design process. They formulate the problem as a reinforcement learning task according to which any candidate algorithm is represented by a policy and the goal is to find an optimal policy. To verify this finding, the authors conducted a set of experiments using different convex and non-convex loss functions correspond to several machine learning models. The obtained results clearly suggest that the automatically designed optimizer converges faster compared to hand-engineered optimizer.

Some of the above mentioned works mainly aim at providing optimal solutions within a very limited computational time [23], while others [3,14] primarily focus on getting better heuristic solutions. These success stories of DL motivated us to investigate the ability of a moderate model so as to make a balance between the solution accuracy and computational time. Altogether, these are the same desired properties in MAs and our work is a step towards investigating the usefulness and strong potential of this research direction.

3 The Proposed Method

This section presents a new optimization method, called MaNet, to explore the possibility of adopting a lightweight deep learning architecture for continuous optimization tasks. In the following, it is assumed that the reader is familiar with the basic concepts of evolutionary computation and deep neural networks.

The MaNet is designed to have the common properties of the MAs: providing a sufficient good solution with incomplete or imperfect information. It starts the optimization procedure with a set of randomly generated solutions as genotype. During training the network, MaNet applies the network training components directly on the genotype, while decodes a genotype into a phenotype (i.e., individuals in MAs) only in the last layer. It finds an optimized solution by iteratively improving an initial solution with regard to its cost function. Among different DL models, CNNs trained with an extension of stochastic gradient descent is used to build the MaNet. The CNNs have been central to the largest advances in computer vision [12] and speech processing [8]. A CNN is a DL method that uses convolutional layers to filter redundant or even irrelevant input data to increase the performance of the network [7]. This consideration also reduces the dimensionality of the input data and speeds up the learning process in the CNNs. Besides, it allows CNNs to be deeper networks with fewer parameters. Altogether, these properties could make CNNs a potential tool for solving optimization problems; especially when we take into account the history behind the application of feature selection [17] and problem scale reducing [21] in the optimization domain.

The architecture of a CNN consists of an input and an output layer, as well as one or more hidden layers. The hidden layers are typically composed of convolutional layers, fully connected layers, normalization layers and pooling layers. The number of hidden layers could be increased depending on the complexities in the input data, but at the cost of more computational expensive simulations. From the mathematical perspective, convolution layers provide a way of mixing input data with a filter so as to form a transformed feature map. Fully-Connected layers learn non-linear combinations of the high-level features by connecting neurons in one layer to neurons in the previous layer, as seen in multi-layer perceptrons neural networks (MLPs). Moreover, normalization layers are adopted to normalize the data to a network and to speed up learning. This includes batch normalization [20], weight normalization [19], and layer normalization [13] techniques. Batch normalization is applied to the input data or to the activation of a prior layer, weight normalization is applied to the weights of the layer and layer normalization is applied across the features. The pooling layers are usually inserted in-between successive convolutional layers to further reduce the number of parameters in the network. A CNN network can have local or global pooling layers that may compute a max or an average.

Inspired by the aforementioned components in CNNs, the MaNet is designed to train a model so as to solve an optimization problem (Fig. 1). The existing feature selection and dimensionality reduction policies in CNNs help MaNet to find complex dependencies between the parameters. The MaNet start optimization by generating a set of random $n \times m$ inputs for the model (i.e., the raw pixel values of the image). So, each individual solution is represented by a matrix rather than a vector. During training the network, convolutional layers transform the initial population layer by layer to a final feasible solution. This large part genotype representation enables the optimizer to keep genetic information that was necessary in the past as a source of exploration, as well as a playground for extracting new features that can be advantageous in the exploitation.

The MaNet multiplies the initial population with a two-dimensional array of filters that are connected to every disjoint region. The output of multiplying the filters with initial population forms a two-dimensional output array called as "feature map". They are obtained by convolution process upon the initial population with a linear filter, without applying a non-linear function or applying feature normalization methods. Similar to other DL models, the filters/kernels in MaNet are learned using the back-propagation algorithm for each specific optimization task. This is the novel aspect of DL techniques that filter weights are learned during the training of the network and are not hand designed. Accordingly, CNNs are not limited to image data and could be used to extract a variety types of features. Thank to this characteristic, MaNet will be forced to extract the features that are the most important to minimize the loss function for the problem at hand the network is being trained to solve. In each convolution layer, we have some predefined hyperparameters that can be used to modify the behavior of the model: the filter size and the number of filters. The first one simply denotes the dimensions of the filter when applying the convolution process, while the second one determines the number of different convolution filters.

In MaNet, multiple convolution layers are stacked which allows convolution layers to be applied to the output of the previous layer, results in a hierarchically set of more decomposed features. Finally, a Dense layer (or fully-connected) with linear activation function will be used to form the final solution vector. As it can be seen from Fig. 1, MaNet has a very simple structure and can benefit from the advantage of having a fast network training process[1]. Indeed, it has only 3,742 trainable parameters compared to state-of-the-art models [22] which

---

[1] Netron Visualizer is used to illustrate the model. The tools is available online at: https://github.com/lutzroeder/netron.

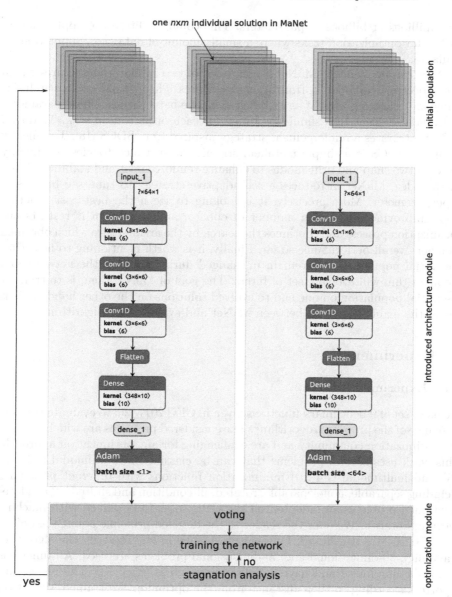

Fig. 1. An overview of the proposed optimization architecture. The MaNet is composed of three convolution layers and one Dense layer (or fully connected layer). In each layer, the number of filters and the filter size are 6 and 3, respectively. The activation function for all the layers is proportional to their inputs.

have millions or billions of parameters. This could facilitate the application of MaNet for optimization tasks where a small amount of data (i.e., population) is available.

As it can be seen, the MaNet is composed of two similar architectures which are subjected to different optimization procedures. The first one uses a batch size of one and the other uses 64 as its batch size. The batch size is a hyperparameter of gradient descent that should be tuned for each optimization task. To do so, MaNet integrates a reinforcement strategy inspired from SDCS [18]. Technically speaking, SDCS is a simple metaheuristic algorithm which toggles continually between two snap and drift modes to enhance reinforcement and stability. Based on this idea, MaNet introduces a self-adaptive strategy to tune the batch size hyperparameter. More precisely, it is looking to see if the best cost function stops improving after some number of epochs, and if so then it restarts the optimization process and continues the search by the architecture which obtained a higher overall performance so far. Finally, it is worth mentioning to note that the initial population will remain unchanged during training the network and the algorithm will evolve a set of filters. The goal of MaNet then, is to transfer the initial population on one end to evolved solutions on the other hand. This is one of the main differences between MaNet and evolutionary algorithms.

4   Experimental

4.1   Experimental Setup

We use a set of 9 benchmark functions given in CEC 2017 [20] to evaluate the performance of the proposed algorithm[2]. The considered problems are widely used in the optimization community and are challenging for any optimization approach. This work uses several problems that can be classified into unimodal (F1 and F3) and multimodal (F4–10) minimization functions with different properties including separable, non-separable, rotated, ill-condition and shifted[3]. The aforementioned problems are adopted on the GPU so as to be linked with machine learning libraries. We refer the reader to the detailed principle about the definition of CEC2017 benchmark functions as defined in [4]. To verify the algorithm scalability, 30-dimensional and 50-dimensional problems are used. All functions should be minimized and have a global minimum at $f(x) = 0$. The results are reported according to their distance from the optimum. We trained MaNet on each problem by using the parallel power of 9 NVIDIA Tesla K20m GPU cards.

It has been shown that various extensions of the differential evolution (DE) [24] algorithm are always among the winners of the CEC competition. Having this is mind, we used jSO [6] algorithm for the purpose of comparison which is the second ranked algorithm in CEC2017 competitions for the single objective

---

[2] The codes for CEC problems and the jSO algorithm are publicly available at: http://www.ntu.edu.sg/home/EPNSugan/index_files/CEC2017/CEC2017.htm.

[3] F2 has been excluded by the organizers because it shows unstable behavior especially for higher dimensions [4].

optimization track. The algorithm is shown to outperform LSHADE [26] (the winner of the CEC2014) and its new extension for CEC2016 (iL-SHADE [5]). All the results are taken from the original study. In order to make a fair comparison, all the experiment conditions are the same. The number of function evaluations is $10,000 \times D$, where D is the problem dimension [4]. To tackle the negative effects of the random initial configurations, each algorithm were run 51 times [4]. The initial population is generated randomly within the search bounds $[-100, 100]$. The parameters of the jSO are the same as reported in the original study [6]. In MaNet, we have 3 convolution layers which are sequentially connected to each other. In each layer, the number of filters and the filter size are 6 and 3, respectively. The MaNet is a CNN model and needs a lot of input data to be well trained and so the population size is fixed to $n = 5,000$. Moreover, m is considered to be 64 for all the problems. The MaNet will be optimized using the Adam algorithm [11].

4.2   Results and Discussion

Tables 1 and 2 present best, worst, mean and standard deviation (Std.) results of the MaNet and jSO on 9 problems over 51 runs. Table 1 reports the results for 30 dimensional problems, while Table 2 shows the performance of the competitive algorithms for 50 dimensional cases. In these tables, a statistical test is also presented to assess the significance of performance between the results of the jSO and MaNet.

Table 1. The obtained results by MaNet and jSO for 30 dimensional problems over 51 runs [4]. The results for jSO are directly taken from the original paper [6].

Function	Algorithm	Best	Worst	Mean	Median	Std.	Sign
1	MaNet	3.71e+02	1.33e+03	7.94e+02	8.02e+02	2.03e+02	−
	jSO	0.00e+00	0.00e+00	**0.00e+00**	0.00e+00	0.00e+00	
3	MaNet	3.69e+04	7.10e+04	5.85e+04	5.85e+04	6.46e+03	−
	jSO	0.00e+00	0.00e+00	**0.00e+00**	0.00e+00	0.00e+00	
4	MaNet	1.46e−05	3.99e+00	**5.88e−01**	6.79e−04	1.41e+00	+
	jSO	5.86e+01	6.41e+01	5.87e+01	5.86e+01	7.78e−01	
5	MaNet	0.00e+00	1.99e+00	**5.85e−01**	1.34e−07	6.59e−01	+
	jSO	3.98e+00	1.32e+01	8.56e+00	8.02e+00	2.10e+00	
6	MaNet	0.00e+00	0.00e+00	**0.00e+00**	0.00e+00	0.00e+00	=
	jSO	0.00e+00	0.00e+00	**0.00e+00**	0.00e+00	0.00e+00	
7	MaNet	3.26e+01	3.41e+01	**3.33e+01**	3.33e+01	3.91e−01	+
	jSO	3.61e+01	4.31e+01	3.89e+01	3.91e+01	1.46e+00	
8	MaNet	0.00e+00	4.97e+00	**2.29e+00**	1.00e+00	1.15e+00	+
	jSO	4.97e+00	1.30e+01	9.09e+00	8.96e+00	1.84e+00	
9	MaNet	0.00e+00	0.00e+00	**0.00e+00**	0.00e+00	0.00e+00	=
	jSO	0.00e+00	0.00e+00	**0.00e+00**	0.00e+00	0.00e+00	
10	MaNet	1.09e+04	1.13e+04	1.11e+04	1.11e+04	1.19e+02	−
	jSO	1.04e+03	2.04e+03	**1.53e+03**	1.49e+03	2.77e+02	

Table 2. The obtained results by MaNet and jSO for 50 dimensional problems over 51 runs [4]. The results for jSO are directly taken from the original paper [6].

Function	Algorithm	Best	Worst	Mean	Median	Std.	Sign
1	MaNet	3.67e+02	2.06e+03	1.39e+03	1.46e+03	3.71e+02	−
	jSO	0.00e+00	0.00e+00	**0.00e+00**	0.00e+00	0.00e+00	
3	MaNet	9.80e+04	1.42e+05	1.23e+05	1.25e+05	8.88e+03	−
	jSO	0.00e+00	0.00e+00	**0.00e+00**	0.00e+00	0.00e+00	
4	MaNet	3.10e−06	1.53e−03	**8.22e−04**	9.96e−04	4.46e−04	+
	jSO	1.32e−04	1.42e+02	5.62e+01	2.85e+01	4.88e+01	
5	MaNet	1.99e+00	1.09e+01	**6.15e+00**	5.97e+00	2.20e+00	+
	jSO	8.96e+00	2.39e+01	1.64e+01	1.62e+01	3.46e+00	
6	MaNet	0.00e+00	0.00e+00	**0.00e+00**	0.00e+00	0.00e+00	=
	jSO	0.00e+00	0.00e+00	**0.00e+00**	0.00e+00	0.00e+00	
7	MaNet	5.49e+01	5.65e+01	**5.58e+01**	5.59e+01	3.62e−01	+
	jSO	5.75e+01	7.42e+01	6.65e+01	6.66e+01	3.47e+00	
8	MaNet	1.99e+00	8.95e+00	**5.41e+00**	5.97e+00	1.99e+00	+
	jSO	9.95e+00	2.41e+01	1.70e+01	1.70e+01	3.14e+00	
9	MaNet	0.00e+00	0.00e+00	**0.00e+00**	0.00e+00	0.00e+00	=
	jSO	0.00e+00	0.00e+00	**0.00e+00**	0.00e+00	0.00e+00	
10	MaNet	1.86e+04	1.88e+04	1.87e+04	1.87e+04	6.25e+01	−
	jSO	2.40e+03	3.79e+03	**3.14e+03**	3.23e+03	3.67e+02	

The results of the Wilcoxon rank sum test are reported at the 95% confidence level. In these tables, '+' shows that MaNet significantly outperforms the jSO with 95% certainty; '−' indicates that the jSO is significantly better than MaNet; and '=' shows there is no statistical different between the two compared algorithms. The significant results are given in bold. For further validation, convergence graphs of jSO and MaNet for 30 dimensional functions F4 and F8 are given in Fig. 2.

As can be seen from Tables 1 and 2, jSO gives more accurate solutions for the unimodal benchmarks F1 and F3 for both 30-dimensional and 50-dimensional cases. Moreover, with the exceptions of F10, MaNet has equal or significantly better performance on all the multimodal benchmark functions. In fact, the results indicate that MaNet significantly outperforms the jSO on 4 functions (F4–F8), obtains an equal performance on 2 functions (F6 and F9), and has worst results on 3 test cases (F1, F3 and F10). Furthermore, we can see that MaNet is a robust algorithm according to the reported standard deviation results. In addition, these experimental results have confirmed that MaNet is not very sensitive to the increment of dimension and is scalable. Considering Fig. 2, it can be seen also that MaNet has a more rapid convergence rate than the jSO algorithm for function F4 and F8. In MaNet, we assume that not selection, but rather the combination of different filters is the main source of evolution and that is the reason for having unstable convergence behavior on these functions.

Altogether, these promising results have confirmed that MaNet has a competitive results in comparison with one of the best designed algorithm for the

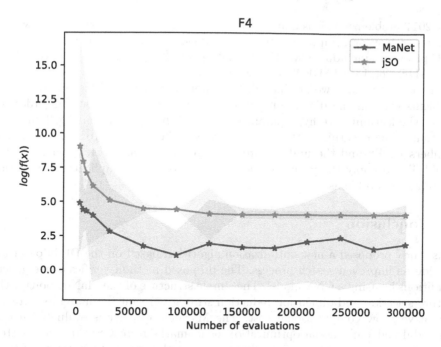

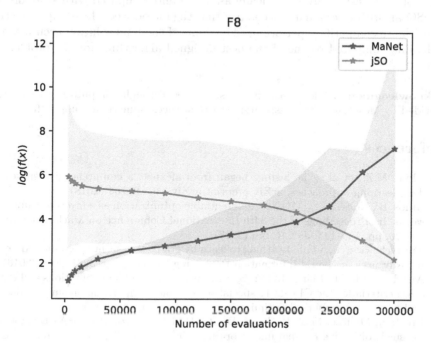

Fig. 2. Convergence graphs of the jSO and MaNet for 30 dimensional functions F4 and F8 over 51 runs

CEC2017 problems. This is quite interesting because MaNet doesn't borrow any search strategy or components from the previously proposed methods for the CEC problems; including CMAES [15], DE, jADE [27], SADE [16], SHADE [25], L-SHADE [26], i-LSHADE [5] and jSO.

As a future work, we are intended to apply the proposed MaNet to all the problems over all the dimensions. Besides, we have to find a way in order to adjust the learning rate hyperparameter for each problem. From Fig. 2 one can see that a high learning rate in Adam causes the network to generate large numbers for F8 and the updates are going to be just as large. After that, we would like to apply the proposed methodology to more complicated real-world optimization problems.

5 Conclusion

This study proposed a new optimization algorithm based on the DL in order to provide an improved search process. The proposed method verifies convergence conditions by using a CNN model. The simple structure of the MaNet along with feature selection and dimension reduction strategies result in an architecture at a relatively low computational cost. The MaNet optimizer is evaluated using unimodal and multimodal optimization benchmarks from CEC2017 test suite. The obtained results are statistically analyzed and compared with state-of-the-art jSO algorithm. Evaluations confirm that the introduced MaNet optimization model has a competitive performance in terms of the final solution accuracy and scalability compared to one of the best designed algorithms for the problem at hand.

Acknowledgments. This research was supported through computational resources provided by Mésocentre of Strasbourg: https://services-numeriques.unistra.fr/.

References

1. Alom, M.Z., et al.: The history began from alexnet: a comprehensive survey on deep learning approaches. arXiv preprint arXiv:1803.01164 (2018)
2. Amos, B., Kolter, J.Z.: OptNet: differentiable optimization as a layer in neural networks. In: Proceedings of the 34th International Conference on Machine Learning, vol. 70, pp. 136–145. JMLR. org (2017)
3. Andrychowicz, M., et al.: Learning to learn by gradient descent by gradient descent. In: Advances in Neural Information Processing Systems, pp. 3981–3989 (2016)
4. Awad, N., Ali, M., Liang, J., Qu, B., Suganthan, P.: Problem definitions and evaluation criteria for the CEC 2017 special session and competition on single objective real-parameter numerical optimization. Technical report (2016)
5. Brest, J., Maučec, M.S., Bošković, B.: iL-SHADE: improved L-SHADE algorithm for single objective real-parameter optimization. In: 2016 IEEE Congress on Evolutionary Computation (CEC). pp. 1188–1195. IEEE (2016)
6. Brest, J., Maučec, M.S., Bošković, B.: Single objective real-parameter optimization: algorithm JSO. In: 2017 IEEE Congress on Evolutionary Computation (CEC), pp. 1311–1318. IEEE (2017)

7. He, K., Zhang, X., Ren, S., Sun, J.: Deep residual learning for image recognition. In: Proceedings of the IEEE Conference on Computer Vision and Pattern Recognition, pp. 770–778 (2016)
8. Hinton, G., et al.: Deep neural networks for acoustic modeling in speech recognition. Signal Process. Mag. **29**, 82–97 (2012)
9. Kang, K., Bae, C., Yeung, H.W.F., Chung, Y.Y.: A hybrid gravitational search algorithm with swarm intelligence and deep convolutional feature for object tracking optimization. Appl. Soft Comput. **66**, 319–329 (2018)
10. Kennedy, M.P., Chua, L.O.: Neural networks for nonlinear programming. IEEE Trans. Circuits Syst. **35**(5), 554–562 (1988)
11. Kingma, D.P., Ba, J.: Adam: a method for stochastic optimization. arXiv preprint arXiv:1412.6980 (2014)
12. Krizhevsky, A., Sutskever, I., Hinton, G.E.: Imagenet classification with deep convolutional neural networks. In: Advances in Neural Information Processing Systems, pp. 1097–1105 (2012)
13. Lei Ba, J., Kiros, J.R., Hinton, G.E.: Layer normalization. arXiv preprint arXiv:1607.06450 (2016)
14. Li, K., Malik, J.: Learning to optimize. arXiv preprint arXiv:1606.01885 (2016)
15. Loshchilov, I.: CMA-ES with restarts for solving CEC 2013 benchmark problems. In: 2013 IEEE Congress on Evolutionary Computation, pp. 369–376. IEEE (2013)
16. Qin, A.K., Suganthan, P.N.: Self-adaptive differential evolution algorithm for numerical optimization. In: 2005 IEEE Congress on Evolutionary Computation, vol. 2, pp. 1785–1791. IEEE (2005)
17. Rakhshani, H., Idoumghar, L., Lepagnot, J., Brévilliers, M.: MAC: many-objective automatic algorithm configuration. In: Deb, K., et al. (eds.) EMO 2019. LNCS, vol. 11411, pp. 241–253. Springer, Cham (2019). https://doi.org/10.1007/978-3-030-12598-1_20
18. Rakhshani, H., Rahati, A.: Snap-drift cuckoo search: a novel cuckoo search optimization algorithm. Appl. Soft Comput. **52**, 771–794 (2017)
19. Salimans, T., Kingma, D.P.: Weight normalization: a simple reparameterization to accelerate training of deep neural networks. In: Advances in Neural Information Processing Systems, pp. 901–909 (2016)
20. Santurkar, S., Tsipras, D., Ilyas, A., Madry, A.: How does batch normalization help optimization? In: Advances in Neural Information Processing Systems, pp. 2483–2493 (2018)
21. Senjyu, T., Saber, A., Miyagi, T., Shimabukuro, K., Urasaki, N., Funabashi, T.: Fast technique for unit commitment by genetic algorithm based on unit clustering. IEE Proc.-Gener. Transm. Distrib. **152**(5), 705–713 (2005)
22. Simonyan, K., Zisserman, A.: Very deep convolutional networks for large-scale image recognition. arXiv preprint arXiv:1409.1556 (2014)
23. Snoek, J., et al.: Scalable Bayesian optimization using deep neural networks. In: International Conference on Machine Learning, pp. 2171–2180 (2015)
24. Storn, R., Price, K.: Differential evolution-a simple and efficient heuristic for global optimization over continuous spaces. J. Glob. Optim. **11**(4), 341–359 (1997)
25. Tanabe, R., Fukunaga, A.: Success-history based parameter adaptation for differential evolution. In: 2013 IEEE Congress on Evolutionary Computation, pp. 71–78. IEEE (2013)

26. Tanabe, R., Fukunaga, A.S.: Improving the search performance of shade using linear population size reduction. In: 2014 IEEE Congress on Evolutionary Computation (CEC), pp. 1658–1665. IEEE (2014)
27. Zhang, J., Sanderson, A.C.: JADE: adaptive differential evolution with optional external archive. IEEE Trans. Evol. Comput. **13**(5), 945–958 (2009)

Evolving a Weighted Combination of Text Similarities for Authorship Attribution

Youssef Keyrouz[1,2]($\boxtimes$) (ID), Cyril Fonlupt[1] (ID), Dany Mezher[2], Denis Robilliard[1], and Rafic Faddoul[2]

[1] Université Du Littoral Côte D'opale, 1 Place de l'Yser BP 71022, 59375 Dunkerque Cedex 1, France
{youssef.keyrouz,cyril.fonlupt,denis.robilliard}@univ-littoral.fr
[2] Université Saint-Joseph de Beyrouth, sise Rue de Damas, BP 17-5208, Mar Mikhaël, Beyrouth 1104 2020, Lebanon
youssef.keyrouz@net.usj.edu.lb, {dany.mezher,rafic.faddoul}@usj.edu.lb

Abstract. Authorship Attribution (AA) also known as Authorship Identification is the problem of identifying the author of an anonymous text based on its characteristics or features. Among notable features extraction methods used to this end, one can cite, the bag of words methods (BOW) and the semantic and syntactic methods (SSM). BOW methods consider the text as a sequence of tokens and disregard the semantics of the language, whereas SSM rely on advanced natural language processing (NLP) techniques. The features extracted from an anonymous text are compared to features extracted from a corpus of texts written by known authors using several similarity measures. In this paper, we combine multiple results generated using conventional methods (chosen from the literature) and we use a genetic algorithm (GA) to find the optimal weighting distribution. The optimal combination obtained by the GA is then applied, and the author attributed to the anonymous text is selected among a set of known authors based on the highest similarity. The fitness of our GA is the resulting accuracy of the authorship attribution task. A numerical application on a corpus consisting of 3036 books written by 142 authors shows that the proposed method has higher accuracy than conventional methods and achieved satisfying performance.

Keywords: Authorship attribution · Genetic algorithms · Feature extraction · Text similarities

1 Introduction

Authorship attribution (AA) is defined as the problem of identifying the author of an anonymous text, or text whose authorship is in doubt [11]. Stylometry is used to attribute authorship to anonymous or disputed documents. The basics of stylometry were defined by Wincenty Lutosławski in 'Principes de stylométrie' (1890) [13]. Early attempts to quantify the writing style of an author go back to the 19th century, with the research of Mendenhall on the Shakespeare

© Springer Nature Switzerland AG 2020
L. Idoumghar et al. (Eds.): EA 2019, LNCS 12052, pp. 13–27, 2020.
https://doi.org/10.1007/978-3-030-45715-0_2

authorship question (1887) [26]. Mendenhall tried to identify the style of different authors by using the frequency distribution of words of various lengths. The first major advancement in authorship attribution happened in the first half of the 20th century, by Yule (1938; 1944) [27,28] and Zipf [29]. Over the last 60 years, many studies have been conducted to try to solve the problem of authorship attribution. The most influential among them is the work of Mosteller and Wallace (1964) [16] on the "The Federalist Papers", a series of political essays written by Alexander Hamilton, James Madison and John Jay, 12 of which claimed by both James Hamilton and Alexander Madison. They based their study on Bayesian statistics of the frequencies of common words which lead to a discriminating result between the candidate authors. Since then, research in authorship attribution was dominated by attempts to define the best features to extract for the detection of an author style. Nearly 1,000 different measures had been proposed by the end of 1990 [20].

In modern days, the increase of computer processing power and speed as well as the advancements of research in information retrieval, machine learning, and natural language processing (NLP) had a significant impact on authorship attribution. Techniques ranging from statistical approaches like principal component analysis [2] to machine learning and neural networks [6,25] have been studied. The main idea behind computer-based authorship attribution is that the text document should be converted to vectors of numerical values before being processed by statistical or machine learning algorithms. This is referred to as using features extraction methods to generate a vector space model (VSM) [21]. In a VSM, documents are represented as vectors in the space of features. The vector values contain numerical representation of one or more textual features. Comparative studies have been performed to compare many feature extraction methods in terms of accuracy and performance [18,23]. Many methods were described and executed with comparative results between the methods. Those feature extraction methods can be divided into 2 categories: (i) bag of words methods (BOW) and (ii) the semantic and syntactic methods (SSM). BOW methods consider the text as a sequence of tokens and disregard the semantics and the language, whereas SSM considers the semantics and heavily rely on advanced NLP techniques. The most common and basic BOW methods include the generation of the N-Gram representation of a text [4,22]. Other more advanced BOW methods were developed to use neural networks for the purpose of finding a distributed representation of words and phrases based on their usage in a context, resulting in predictive models e.g. 'Word2Vec' [15] or count based models e.g. 'GloVe' [17].

This paper is based on the premise that by generating a linear combination of standard features extraction method results, and then using a genetic algorithm (GA) [24] to find the best linear combination, better accuracy can be obtained. The resulting combination would be more accurate than the individual method results. This comes from the assumption that each method focuses on a different aspect of the author's style. Hence, combining the results and assigning the proper weight for each of them would give a more accurate representation of the overall writing style. The authorship identification can be achieved by applying similarity measures to compare the features of the anonymous texts with features from a corpus of texts having known authors and selecting the best similarity after combining the results.

Section 1 introduces authorship attribution and the related work. Section 2 describes the proposed method for AA and its performance calculation. Section 3 details the experiment on the corpus. Section 4 discusses the results and observations generated from the experiment. And finally, Sect. 5 concludes the study and suggests possible improvements for future work.

2   Method

2.1   Preparing the Data

A textual features extraction method is a process to convert a raw text into numerical features represented by a vector in the VSM. The resulting vector is used for calculations or machine learning. Each method will take a text as an input and provide a multi-dimensional vector as the output. The number of dimensions depends on the method used. i.e. each method has its own representation of vectors and dimensions depending on the feature it extracts. Some methods require the text to be preprocessed before extracting the feature. Preprocessing is performed by applying NLP techniques. In this paper, multiple NLP techniques are implemented to prepare the text before running the features extraction method. Table 1 contains the different ways a text can be represented after preprocessing.

Table 1. Different text representations.

Representation	Description
Raw text	The text is taken as is, without any preprocessing or manipulation
Stemmed text	Each word in the text is stemmed, i.e. replaced by its root form. e.g. words like "cats", "cat", and "kittens" will be replaced by their root form "cat"
Part of Speech (PoS) tagging	Each word in the text is replaced by its type (e.g. pronoun, verb, adjective, adverb $\cdots$)

The methods are applied on a corpus of texts and writings, divided and arranged by author. 25% of the corpus is set aside and considered of unknown authors. They are used to validate the results and calculate the accuracy of the study. To ensure proper results, all authors will have one or multiple books considered unknown and chosen as part of the 25% testing subset. The other 75% are used as the training subset for the study.

2.2   Generating the Profiles

A "Profile" is defined as a numerical vector representing the text of an author. Two different profile approaches are studied. First approach is to have an "Author Profile". This produces one cumulative representation for all training texts per author. In other words, all the books and writings for an author are combined into one big text. Then a profile vector is generated for the entire text. This approach will generate one profile per author and use it for comparison. Second approach is to have an "Individual Profile". Each book or writing is used separately, and a profile vector is generated for each one. The comparison will be performed against each individual profile. In this approach, an author can have many different profiles, considering he or she has as many profiles as books. Table 2 highlights some advantages and disadvantages for each profiling approach.

Table 2. Different profiles and their advantages and disadvantages.

Representation	Advantages	Disadvantages
Author profile	Faster to process due to a lower number of profiles (one profile per author) Can handle the cases where only short texts are available. Their concatenation may produce a more reliable profile	May have a lower accuracy due to the use of multiple texts having different genres or style (e.g. an author who wrote about politics but also a fantasy novel)
Individual profile	Higher accuracy. When comparing against individual books, we have higher chance of matching with a similar book from the same author	Slower to process due to the high number of profiles. Might be inaccurate when dealing with a book having completely different genre than the author's usual writings

2.3   Finding Similarities and Analyzing Performance

The goal is to calculate the similarities between the unknown texts and the profiles from the corpus. A profile is generated for the unknown text and compared with the known profiles, then the best matching profile is selected, and the author of that profile is attributed as the author of the text. To evaluate the performance and accuracy of the attribution method, a list of known profiles is generated from the training subset with known authors $(P_1, P_2, P_3, \ldots)$, and a list of unknown profiles is generated from the testing subset considered of unknown authors $(T_1, T_2, T_3, \ldots)$. After all the profiles are generated, a comparison is made between each unknown profile T and each known profile P by calculating the similarities between T and P, then the best matching profile is identified using nearest-neighbor algorithm [7]. We define the similarity value S_{mn} in Eq. (1).

$$S_{mn} = f(P_m, T_n) \tag{1}$$

Where f is the similarity function, P_m the profile of the known author m (The maximum value of m is the number of profiles in the training set), T_n the profile of an unknown author (The maximum value of n is the number of profiles in the testing set), and S_{mn} the similarity between P_m and T_n.

For a given features extraction method, all the known profiles P and the unknown profiles T are generated. A similarity table is calculated as shown in Table 3. At this stage, each anonymous text (represented by its profile T) will have a similarity value S with each of the known profiles P. The profiles are then sorted from best match to worst match. Each unknown profile T will have a list of possible known profiles P sorted based on the similarity value between them (Best Matching P, 2nd Best Matching P, 3nd Best Matching P,...).

Table 3. Example of a similarity table for a chosen method and a chosen similarity formula, using five profiles and three anonymous texts.

Profiles	P_1	P_2	P_3	P_4	P_5
T_1	S_{11}	S_{21}	S_{31}	S_{41}	S_{51}
T_2	S_{12}	S_{22}	S_{32}	S_{42}	S_{52}
T_3	S_{13}	S_{23}	S_{33}	S_{43}	S_{53}

An example of a result is shown in Table 4. The profile T_1 has the best match with profile P_3, second best match with P_2, then with P_5, then with P_1, and has the worst match with P_4. Same logic is applied for T_2 having the best match with P_5, and T_3 having the best match with P_3.

Table 4. Example of a result after sorting the profiles from best matching to worst matching.

T_1	P_3	P_2	P_5	P_1	P_4
T_2	P_5	P_4	P_1	P_3	P_2
T_3	P_3	P_1	P_5	P_4	P_2

Considering P_1 the profile of a known author 1 (A_1), P_2 the profile of known author 2 (A_2), and so on... T_1 the profile of a text for A_1 but considered unknown, T_2 the profile of a text for A_2 also considered unknown, and so on. The accuracy measure is introduced to measure how well the method is attributing the authors.

Table 5. Example of an authorship attribution output.

Unknown profile	Best matching profile	Attributed author	Real author	Attributed correctly
T_1	P_3	A_3	A_1	NO
T_2	P_5	A_5	A_2	NO
T_3	P_3	A_3	A_3	YES

To calculate the accuracy, the best match is chosen from the similarity table, and the author of the corresponding profile is attributed as the author of the unknown text. A counter will increment every time a correct attribution is made to count the number of correct matches. The accuracy is the percentage of the correct matches. Applying this on the example result from Table 4, the authorship attribution output is shown in Table 5. Only T_3 was attributed correctly which gives an accuracy of: $Acc = (1/3) \times 100 = 33.3\%$.

2.4 Combining the Results and Evolving the Weights Using a Genetic Algorithm

The premise proposed in this section is that by combining the resulting similarity tables obtained from different feature extraction methods, a new similarity table can be generated leading to a higher accuracy of the authorship attribution task. We start with multiple features extraction methods; each method will generate a similarity table (as seen in Table 3). Those tables are combined into one before the ranking is performed. Two approaches are used for the combination.

The first approach is to combine the methods output in a binary way. A method is either used or not. Equation (2) is a binary combination of the similarity tables.

$$M = \sum_{i=1}^{n} b_i M_i = b_1 M_1 + b_2 M_2 + \dots \qquad (2)$$

Where M is the resulting similarity table after the combination, M_i is the similarity table obtained by applying method i from the list of methods to consider, n is the total number of methods, and b_i is the binary multiplier of method i. b_i is equal to 1 if the method is used or 0 if the method is not used.

The second approach is to generate a linear combination. Similar to binary combination but instead of using 1 and 0 as multipliers, a weight between 0 and 1 is assigned to each method and multiplies the similarity table. All methods will contribute to the attribution process but with different weights. The assumption here is that even bad performing methods can have a small impact on the accuracy as it focuses on a specific aspect of the author style. This is achieved by assigning a weight w for each method and doing a linear combination as shown in Eq. (3).

$$M = \sum_{i=1}^{n} w_i M_i = w_1 M_1 + w_2 M_2 + \ldots \qquad (3)$$

Where w_i is the weight assigned to method i. It is a decimal value between 0 and 1. This approach is a fine tuning of the binary combination and gives more precise results.

For both combination approaches, a genetic algorithm (GA) is implemented to find the best binary values combination or weights combination. GA is an adaptive heuristic search algorithm based on the evolutionary ideas of natural selection and genetics. They are used to exploit random search for optimization problems. Each individual in the GA population is a weight distribution for all the methods. A generational Evolution Strategy will be built for the problem requiring the individuals to be vectors of doubles.

To avoid overfitting and have proper accuracy measure, a K-Fold Cross Validation [19] is used on the training data. In k-fold cross-validation, the data is randomly divided into k equal sized subsamples. One of the subsamples is retained as the validation data, and the remaining k − 1 subsamples are used as the training data. This process is then repeated for each subsample (so a total of k times) with each of the subsamples used exactly once as the validation data. After applying the k-Fold Cross Validation, each subsample will have an accuracy. A total of k accuracies are generated and the fitness for an individual is obtained by averaging the accuracy. The fitness is between 0 (worst fitness) and 1 (best fitness).

Putting it all together, each individual is a vector of weights. The weights are assigned to the method outputs to generate their linear combination. This produces a new combination to be used on the training data. After applying a k-fold cross validation, the average accuracy is set as the fitness of the individual. Then the population is evolved based on the Evolution Strategy chosen.

3 Experimentation

3.1 Corpus and Tools

This section will take the method proposed above and apply it on the Gutenberg Dataset [9]. It is a collection of 3036 books and writings for 142 different authors, subset of the Project Gutenberg [8]. All books have been manually cleaned to remove metadata, license information, and transcribers' notes. 25% of the corpus (704 texts) are set aside and considered of unknown authors. The other 75% (2332 texts) are used as the training subset for the study. The method is implemented using the JAVA Programming language. Stanford CoreNLP (Version 3.9.2) [14] provides a set of human language technology tools that are used to preprocess the text, apply stemming, or PoS tagging. ECJ (Version 26) [12], a Java-based evolutionary computation research system, is used to implement the GA and the parameters.

We chose ten conventional features extraction methods from the literature to be used in our experiment. Many comparative studies have been performed on those methods [18,23] and they are used in many papers and studies. The different methods used are listed below.

- **Characters n-Gram frequency**: The text is considered as a contiguous sequence of n characters. This method requires a splitter to split the raw text every n character, creating a collocation of n characters, where the order is also considered. The frequency of each collocation is calculated, and the vector is generated. The resulting numerical vector entries consist of each possible n-gram sequence with the frequency of appearance of that sequence in the text. This paper will include 3 variations of the n-Gram with $n = 1, 2,$ and 3 (Called respectively 1-Gram 2-Gram, and 3-Gram).
- **Words length frequency**: This method calculates the frequency of words with different lengths (i.e. number of characters). The resulting numerical vectors entries consist of each possible length (1 character, 2 characters, 3 characters etc.) and indicate the frequency of words having that length in the text. A tokenizer is needed to extract words from the raw text and count them.
- **Words usage frequency**: This method assumes that authors tend to use special words in their text or rely on a specific word more than others. It counts the frequency of each word in the vocabulary and a numerical vector is generated. Each entry in the vector is a word from the vocabulary represented by its frequency of appearance in the text. This method is applied on a stemmed text and requires a tokenizer to extract the words. Stemming is essential for this method to reduce the dimensionality as well as remove all the noise coming from grammar and different representation of the word (plural, verb conjugations, usage as adjective etc.).
- **Sentence length frequency**: The sentence length frequency counts the frequency of sentences with different word lengths (i.e. number of words in the sentence). A sentence splitter is needed to extract the sentences from the raw text and then a tokenizer is used to count words in each sentence.
- **Commas frequency in sentences**: This method counts the number of commas in the sentence. This relies on the assumption that authors tend to use commas in a distinct way. A sentence splitter is used to extract the sentences from the raw text. Then the commas are counted in each sentence and a numerical vector is generated representing the frequencies of different comma count in the text (e.g. how many sentences have no commas, 1 comma, 2 commas, 3 commas etc.).
- **Verbs usage frequency**: Similar to the "Words usage frequency" described above, but only counting the usage of verbs. This assumes that the authors prefer to use special verbs in their writing. PoS tagging as well as stemming are needed for this method.
- **Verbs frequency in sentences**: This method counts the number of verbs in the sentence. This assumes that some authors tend to use few verbs in a sentence while others prefer to use many verbs in a sentence. PoS tagging must be performed on the text to tag all the verbs before counting.

- **Verbs tenses frequency**: This method counts the usage of verb tenses in the text. This assumes that authors tend to prefer the usage of specific verb tenses. Some examples of verb tenses are: Present perfect, future, simple past, past perfect etc. PoS tagging is performed on the raw text and the verb tenses frequencies are calculated.

This study applies two different similarity formulas for the comparison. The first similarity measure used is the Euclidean Distance. The Euclidean distance is a straight line between two points defined by Eq. (4).

$$S_{mn} = \sqrt{\sum_{i=1}^{l}(P_m^i - T_n^i)^2} \tag{4}$$

Where l the length of the vector generated by the method, P_m^i the value at index i of the known profile m, T_n^i the value at index i of the unknown profile n.

The second similarity measure used is the Cosine Similarity. The cosine similarity is the cosine of the angle formed by two vectors and is defined by Eq. (5).

$$S_{mn} = \frac{\sum_{i=1}^{l}(P_m^i T_n^i)}{\sqrt{\sum_{i=1}^{l}(P_m^i)}\sqrt{\sum_{i=1}^{l}(T_n^i)}} \tag{5}$$

The logic to choose the best match depends on which similarity measure is used. Table 6 shows the logic of choosing the best match for each similarity measure. Optimizing the evolution strategy and the parameters of the GA is out of the scope of this study but will be considered in future works. A commonly used evolution strategy, the mu+lambda evolution strategy [1], is chosen with default parameters to evolve a vector of doubles. The GA is configured with a mutation probability of 0.2 using a gaussian convolution on the values with a standard deviation of 0.3 and evolved over 1000 generations.

Table 6. Choosing the best match for each similarity measure.

Similarity measure	Best match
Euclidean distance	The best match is the value closest to 0. And since all the values are positive, the best match corresponds to the lowest value in the row
Cosine similarity	The best match is the value closest to 1 indicating the vectors are equal. And since all the values are between −1 and 1, the best match corresponds to the highest value in the row

4   Results and Discussions

Figure 1 compares the accuracy of each method for both the individual profile approach and the author profile approach using the Euclidean distance. Figure 2 shows the same comparison but using the Cosine similarity.

A clear difference in accuracy is visible between the methods. Using the Euclidean distance for the individual profile approach, the highest two methods are the "2-Gram" and "3-Gram" methods with an accuracy of 80% and 81% respectively. Accuracy goes down to 64% for the "2-Gram" and 66% for the

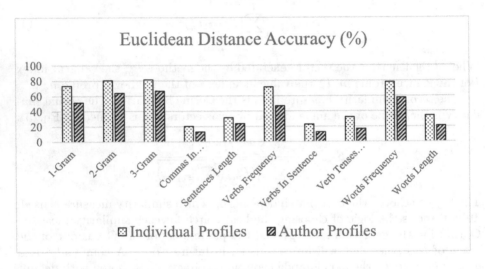

Fig. 1. Methods accuracy using the Euclidean distance.

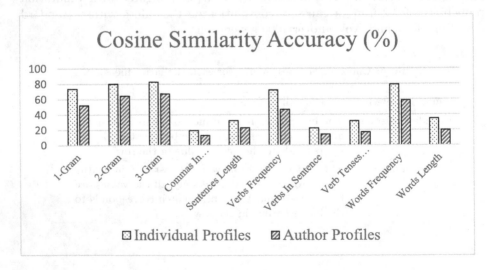

Fig. 2. Methods accuracy using the Cosine similarity.

"3-Gram" methods when using the author profile approach. Some methods have very low accuracy, e.g. the "Commas in Sentence" method yielding an accuracy of 20% for the individual profile and 13% for the author profile using the same Euclidean distance. Some interesting observations can be made from the results shown in Figs. 1 and 2.

- The Euclidean distance and the Cosine similarity give almost similar accuracy and pattern. This shows that the accuracy of the method is independent of the similarity measure used.
- The individual profile approach has better accuracy than the author profile. This proves the assumption that it is more likely to match a book with a similar book from the same author, than matching a book with the entire author bibliography due to the noise introduced by having books of different styles.

Next step is evolving the genetic algorithm to improve the fitness of our individuals. Figure 3 shows how the fitness evolved over 1000 generations (Chosen as the stopping criterion).

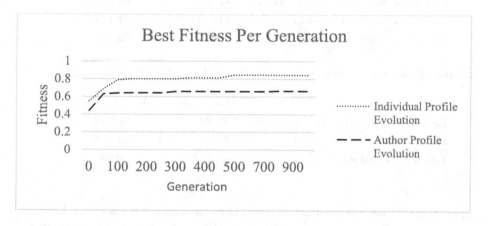

Fig. 3. Evolving the linear weights combination using the Euclidean Distance.

Table 7 contains the best weights chosen genetically. Observations on the weights in Table 7:

- The binary combination approach resulted in the GA choosing the two best methods in combination to improve the results. Adding any more methods will cause a loss of accuracy due to the addition of noise.
- The linear combination approach resulted in a fine tuning of that choice. We can clearly see that the two methods chosen in the binary combination approach have the biggest weights assigned to them. Methods that are less performing have lower weights assigned to them.

Table 7. Best methods combination weights after genetic evolution.

Method	Binary combination	Linear combination weights
1-Gram	0	0.83
2-Gram	1	0.95
3-Gram	1	0.99
Commas in sentence	0	0.03
Sentences length	0	0.06
Verbs frequency	0	0.33
Verbs in sentence	0	0.5
Verb tenses frequency	0	0.032
Words frequency	0	0.53
Words length	0	0.05

Table 8. Accuracy before and after genetic evolution for the Euclidean distance.

	Individual profile	Author profile
Best single method accuracy (3-Gram)	81.38%	66.62%
Binary combination accuracy	84.60%	67.41%
Linear combination accuracy	86.80%	70.30%

Table 9. Accuracy before and after genetic evolution for the Cosine Similarity.

	Individual profile	Author profile
Best single method accuracy (3-Gram)	82.37%	67.33%
Binary combination accuracy	85.10%	67.95%
Linear combination accuracy	86.91%	70.20%

Tables 8 and 9 show the accuracy changes between a single method (the best one was chosen for the comparison), combining the methods in a binary approach, and combining the methods in a linear approach. This accuracy is calculated on the test dataset of unknown authors.

Further evaluations are performed to study the effect of varying the number of authors in the corpus on the performance of the attribution. Table 10 shows the results of the accuracy improvement between a single method, a binary combination, and a linear combination when changing the number of authors in the corpus for the individual profile using the Euclidean distance. Subsets of 5, 20, and 50 authors are randomly selected from the corpus, and the attribution accuracy calculated for each subset. To avoid bias in the results, the author selection process is performed multiple times, randomizing the authors in the subset on each trial. The results shown in Table 10 consist of the average accuracy for all the trials.

Table 10. Attribution accuracy of the individual profile and Euclidean distance for a varying number of authors.

	5 Authors	20 Authors	50 Authors	142 Authors
Best single method accuracy	90.7%	88.9%	84.7%	81.4%
Binary combination accuracy	91.0%	90.1%	86.9%	84.6%
Linear combination accuracy	95.8%	91.9%	87.7%	86.8%

Combining method results improved the accuracy of the attribution. Fine tuning it by applying a linear combination improved the accuracy even more. This shows that each method has a contribution to make but should have the proper weight assigned to that contribution.

5 Conclusion

In this paper, we have presented an approach to combine multiple features selection method results and used Genetic Algorithm to find the optimal weighting distribution for the methods. The method proposed can be considered as a simple Hyper-Heuristic [3] approach. We also studied two different profiling approaches; The individual profile approach and the author profile approach. A numerical application on a corpus consisting of 3036 books written by 142 authors using two different similarity functions has shown that the presented approach improves the performance of the authorship attribution.

Future work on this subject could improve it even further by choosing a more performing evolution strategy. An automatic algorithm configuration e.g. "The irace package" [10] can be used to find the best parameter settings. More advanced methods can be studied and used to extract better features and then added to the method combination. More advanced Hyper-Heuristic approaches can be considered for the combination. And genetic programming [5] can also be used to generate a non-linear combination of the results.

References

1. Beyer, H.G., Schwefel, H.P.: Evolution strategies - a comprehensive introduction. Nat. Comput. **1**(1), 3–52 (2002). https://doi.org/10.1023/a:1015059928466
2. Binongo, J., Smith, M.: The application of principal component analysis to stylometry. Literary Linguist. Comput. **14**(4), 445–466 (1999). https://doi.org/10.1093/llc/14.4.445
3. Burke, E.K., Hyde, M.R., Kendall, G., Ochoa, G., Özcan, F., Woodward, J.R.: A classification of hyper-heuristic approaches: revisited. In: Gendreau, M., Potvin, J.-Y. (eds.) Handbook of Metaheuristics. ISORMS, vol. 272, pp. 453–477. Springer, Cham (2019). https://doi.org/10.1007/978-3-319-91086-4_14
4. Clement, R.: Ngram and Bayesian classification of documents for topic and authorship. Literary Linguist. Comput. **18**(4), 423–447 (2003). https://doi.org/10.1093/llc/18.4.423

5. Day, P., Nandi, A.K.: Evolution of superFeatures through genetic programming. Expert Syst. **28**(2), 167–184 (2010). https://doi.org/10.1111/j.1468-0394.2010.00547.x

6. Ge, Z., Sun, Y., Smith, M.J.T.: Authorship attribution using a neural network language model. In: Proceedings of the Thirtieth AAAI Conference on Artificial Intelligence, AAAI 2916, pp. 4212–4213. AAAI Press (2016). http://dl.acm.org/citation.cfm?id=3016387.3016522

7. Guo, G., Wang, H., Bell, D., Bi, Y., Greer, K.: An kNN model-based approach and its application in text categorization. In: Gelbukh, A. (ed.) CICLing 2004. LNCS, vol. 2945, pp. 559–570. Springer, Heidelberg (2004). https://doi.org/10.1007/978-3-540-24630-5_69

8. Gutenberg: Project gutenberg, March 2018 (n.d.). www.gutenberg.org

9. Lahiri, S.: Complexity of word collocation networks: a preliminary structural analysis. In: Proceedings of the Student Research Workshop at the 14th Conference of the European Chapter of the Association for Computational Linguistics. Association for Computational Linguistics (2014). https://doi.org/10.3115/v1/e14-3011

10. López-Ibáñez, M., Dubois-Lacoste, J., Cáceres, L.P., Birattari, M., Stützle, T.: The irace package: iterated racing for automatic algorithm configuration. Oper. Res. Perspect. **3**, 43–58 (2016). https://doi.org/10.1016/j.orp.2016.09.002

11. Love, H.: Attributing Authorship: An Introduction. Cambridge University Press, Cambridge (2002). https://doi.org/10.1017/CBO9780511483165

12. Luke, S.: ECJ evolutionary computation library. Available for free (1998), http://cs.gmu.edu/~eclab/projects/ecj/

13. Lutoslawski, W.: Principes de stylométrie appliqués à la chronologie des œuvres de platon. Revue des Études Grecques **11**(41), 61–81 (1898). https://doi.org/10.3406/reg.1898.5847

14. Manning, C.D., Surdeanu, M., Bauer, J., Finkel, J., Bethard, S.J., McClosky, D.: The Stanford CoreNLP natural language processing toolkit. In: Association for Computational Linguistics (ACL) System Demonstrations, pp. 55–60 (2014). http://www.aclweb.org/anthology/P/P14/P14-5010

15. Mikolov, T., Chen, K., Corrado, G., Dean, J.: Efficient estimation of word representations in vector space. In: Proceedings of Workshop at ICLR 2013, January 2013

16. Mosteller, F., Wallace, D.L.: Inference in an authorship problem. J. Am. Stat. Assoc. **58**(302), 275 (1963). https://doi.org/10.2307/2283270

17. Pennington, J., Socher, R., Manning, C.D.: Glove: global vectors for word representation. In: Empirical Methods in Natural Language Processing (EMNLP), pp. 1532–1543 (2014). https://doi.org/10.3115/v1/D14-1162. http://www.aclweb.org/anthology/D14-1162

18. Ramezani, R., Sheydaei, N., Kahani, M.: Evaluating the effects of textual features on authorship attribution accuracy. In: ICCKE 2013. IEEE, October 2013. https://doi.org/10.1109/iccke.2013.6682828

19. Refaeilzadeh, P., Tang, L., Liu, H.: Cross-validation. In: Liu, L., Özsu, M.T. (eds.) Encyclopedia of Database Systems, pp. 532–538. Springer, Boston (2009). https://doi.org/10.1007/978-0-387-39940-9_565

20. Rudman, J.: The state of authorship attribution studies: some problems and solutions. Comput. Humanit. **31**, 351–365 (1997). https://doi.org/10.1023/A:1001018624850

21. Salton, G., Wong, A., Yang, C.S.: A vector space model for automatic indexing. Commun. ACM **18**(11), 613–620 (1975). https://doi.org/10.1145/361219.361220

22. Selj, V., Peng, F., Cercone, N., Thomas, C.: N-gram-based author profiles for authorship attribution. In: Proceedings of the Conference Pacific Association for Computational Linguistics PACLING 2003, September 2003
23. Stamatatos, E.: A survey of modern authorship attribution methods. J. Am. Soc. Inf. Sci. Technol. **60**(3), 538–556 (2009). https://doi.org/10.1002/asi.21001
24. Tang, K., Man, K., Kwong, S., He, Q.: Genetic algorithms and their applications. IEEE Signal Process. Mag. **13**(6), 22–37 (1996). https://doi.org/10.1109/79.543973
25. Tweedie, F.J., Singh, S., Holmes, D.I.: Neural network applications in stylometry: the federalist papers. Comput. Humanit. **30**(1), 1–10 (1996). https://doi.org/10.1007/bf00054024
26. Williams, C.B.: Mendenhall's studies of word-length distribution in the works of Shakespeare and Bacon. Biometrika **62**(1), 207–212 (1975). https://doi.org/10.1093/biomet/62.1.207
27. Yule, C.U.: The Statistical Study of Literary Vocabulary. Cambridge University Press, Cambridge (2014)
28. Yule, G.U.: On sentence-length as a statistical characteristic of style in prose: with application to two cases of disputed authorship. Biometrika **30**(3/4), 363 (1939). https://doi.org/10.2307/2332655
29. Zipf, G.K.: Selected Studies of the Principle of Relative Frequency in Language. Harvard University Press, Cambridge (1932). https://doi.org/10.4159/harvard.9780674434929

Image Signal Processor Parameter Tuning with Surrogate-Assisted Particle Swarm Optimization

Geoffrey Portelli and Denis Pallez[✉]

Université Côte d'Azur, CNRS, I3S, Sophia Antipolis, France
geoffreyportelli@gmail.com, denis.pallez@univ-cotedazur.fr

Abstract. Evolutionary algorithms (EA) are developed and compared based on well defined benchmark problems, but their application to real-world problems is still challenging. In image processing, EA have been used to tune a particular image filter or in the design of filters themselves. But nowadays in digital cameras, the image sensor captures a raw image that is then processed by an Image Signal Processor (ISP) where several transformations or filters are sequentially applied in order to enhance the final picture. Each of these steps have several parameters and their tuning require lot of resources that are usually performed by human experts based on metrics to assess the quality of the final image. This can be considered as an expensive black-box optimization problem with many parameters and many quality metrics. In this paper, we investigate the use of EA in the context of ISP parameter tuning with the aim of raw image enhancement.

Keywords: Image Signal Processor · Parameter tuning · Particle Swarm Optimization

1 Introduction

Image processing has been largely investigated for many years. As the quality of outputs is crucial for the whole computer vision chain, sophisticated mathematical theories and statistical methods have been developed in recent years [7]. These are a source of complex optimization problems. Moreover, new constraints for embedded, real-time computer vision systems necessitate the development of robust and flexible as well as cost-effective algorithms.

Image-related optimization problems tend to be highly nonlinear and computationally expensive. The objectives and constraints are usually related to image metrics that need to be calculated based on the images from the output of the whole computer vision chain. Thus each evaluation of such objectives requires the combination of applying the considered image processing techniques (denoising, sharpening, white balance, etc...) with optimization-related techniques. Moreover the computational costs will increase with the size of the images. All these factors lead to high computational costs [18].

© Springer Nature Switzerland AG 2020
L. Idoumghar et al. (Eds.): EA 2019, LNCS 12052, pp. 28–41, 2020.
https://doi.org/10.1007/978-3-030-45715-0_3

Classical image processing methods tuned with EAs have been shown to outperform manually tuned classical methods such as image filtering, denoising or enhancement [17], and object detection [5], in terms of convergence speed and ability to deal with highly degraded image source. However, because of its complexity (non-linear process, number of parameters, etc...) and computing resource needs, authors focused on the optimization of a unique task [3,12,18]. For instance, [3] investigated the application of the Artificial Bee Colony algorithm (ABC) [10] where the artificial bees are moving to search for the optimal parameters of a pixel luminance transformation function to enhance the image contrast based on the fitness function.

But nowadays, embedded computer vision systems can be found in lots of devices such as digital cameras and small devices as smartphones, tablets, etc. Usually coupled with the digital sensor, an ISP can be found and applies sequentially many processing and filtering to the raw images like noise reduction, sharpening, white balance or color correction and thus aims at enhancing the image quality of final pictures (Fig. 1). Depending on the application, ISPs can have approximately 6 to 14 processing stages and each stage can have tens or even hundreds of parameters.

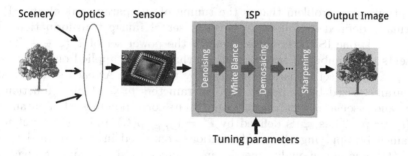

Fig. 1. Classical design of a digital vision system.

All those processing are based on methods that are parameterized. When enhancing only one raw image, many parameters have to be set and each of them could be continuous or discrete. Because of its size and its complexity, i.e. the amount of parameters to optimize (constrained and unconstrained) and several time consuming criteria, the problem to be solved combines fundamental issues like:

– large scale optimization: each stage embedded in the ISP can have few or even hundreds of parameters,
– constrained optimization: processing stages might have parameters that must fulfill specific inequality to be valid, for instance a specific relationship between two (or more) parameters or parameter values that must be within a given range,

- many objectives: several metrics can be used to asses the quality of the output image,
- expensive optimization: computational costs increase with the image size and ISP may need several minutes to process one raw image.

In addition, a gradient based search is not usable because the problem to optimize is a set of methods or procedures that might not be differentiable.

Usually, the parameters of ISPs are painstakingly tuned by a human expert based on image quality metrics. This painstaking work usually leads to one close-to-optimal solution and must be done for each product and variant, severely limiting the number of new configurations that can be evaluated.

Although EAs have been applied to image processing considering one unique task as showed previously, to our knowledge EAs have never been used to optimize a whole image processing chain made of several sequential processing stages as ISPs. Thus here, we investigate the usage of such EA to automatically optimize parameters of a typical ISP based on image quality metrics, for the enhancement of raw images.

2   Real-World Problem: ISP

The optimization problem that is the tuning of the parameters of an ISP can be formally defined as following. Given a set of image transformations $\mathbb{T} = \{t_1, t_2, ..., t_k\}$, one ISP φ is one element of the power set of $\mathbb{T}$ ($\varphi \in \mathcal{P}(\mathbb{T})$). $|\varphi|$ represents the number of transformations that will be applied on a raw image I_R by φ. $\mathbb{P}$ is the set of parameters used by all $t_i \in \varphi$. As each transformation t_i is parameterized by a specific set of parameters $p_i \subseteq \mathcal{P}(\mathbb{P})$, a function p is defined and associates a transformation with its corresponding set of parameters, i.e. $p(t_i) = p_i$. Thus, φ is defined by $\mathbb{P}' = \bigcup_{t_i \in \varphi} p(t_i)$ and, the output image I obtained by applying all transformations contained in φ is defined by $I = \varphi(I_R, \mathbb{P}')$. Some image quality metrics m_i are measured on resulting image I.

The goal is to identify the parameter values of $\mathbb{P}'$ that will optimize all m_i metrics. Without any loss of generality, this optimization problem can be formalized by Eq. 1:

$$\arg\min_{x \in \mathbb{R}^{|\mathbb{P}'|}} \quad M(x) = \{m_1(x), \ldots, m_n(x)\}$$

$$\text{subject to} \quad g_i(x) \leq 0, \quad i = 0, \ldots, m \quad m \geq 0 \tag{1}$$

$$h_j(x) = 0, \quad j = 0, \ldots, p \quad p \geq 0$$

where g_i and h_i are respectively inequality and equality constraints on transformations parameters.

3   Objectives: Image Quality Metrics

In the domain of camera-equipped mobile devices, a standardized suite of objective and subjective image quality metrics is available [1]. This suite also provides standardized image targets that allow the metrics to be comparable between different camera devices. Here are considered the visual noise and the visual acutance (as defined in [1]) to assess the noise and the acutance of the image, respectively. Indeed, a human observer with little experience can visually perceive if an image subjectively looks soft or sharp, also if it looks really noisy or not. Thus, a visual validation of solutions given by an EA can be done contrary to more "abstract" metrics such as Root Mean Square Error or Signal over Noise Ratio that are informative but difficult to sense with human eyes, for instance.

The image noise can be view as the random variation of brightness or color information in images, usually related to the electronic noise. It can be produced by the digital sensor and circuitry in digital camera. The acutance of an image describes a subjective perception of sharpness that is related to the edge contrast of an image. It is linked to the amplitude of the derivative of brightness with respect to space.

The output value of visual noise function will actually decrease as the noise decrease in the measured image. Inversely, the output of the visual acutance function will increase with the sharpness of the measured image: the more the sharpness, the greater the visual acutance. One has to note that noise and acutance counteract with each other.

From a human point of view, a high value of visual acutance does not necessarily implies a subjectively good looking image. Preliminary tests with manual tuning of the ISP used here (data not shown) show that values of visual acutance $\sim$102.0 (and visual noise $\sim$0.01) gives best looking images. Thus not the raw value of the visual acutance function but its transformation $(x - 102.0)^2$ will be used as objective to be minimized. The visual noise as it is will be minimized.

4   Particle Swarm Optimization

The aim of the presented study is not to compare EAs nor to find the best EA for the considered problem. Here the aim is to test whether or not this kind of real-world problem can be tackled using EA. Among the numerous EA strategies that have been investigated by the community, here in the considered real-world application, we focus on Particle Swarm Optimization algorithms (PSO). However others EA strategies could be considered.

PSO is a bio-inspired meta-heuristic mimicking the social behavior of bird flocking or fish schooling [11] which has become very popular to solve multi-objective [15] and many-objective [6] optimization problems. Briefly, swarm particles will evolve in the solution space where their position will be updated at each iteration given their own vector speed computed from global best and local best encountered solutions.

PSO has many advantages, for instance, easy implementation, effective memory, efficient maintenance of the solution diversity [19]. Also, PSO makes few or no assumptions about the problem being optimized and can search very large spaces of candidate solutions. PSO does not use the gradient of the problem being optimized, which means PSO does not require that the optimization problem be differentiable as is required by classic optimization methods such as gradient descent. However, as many others metaheuristics, PSO do not guarantee an optimal solution is ever found.

Among all the possible variants, we focus on SMPSO [14] to solve the optimization of the ISP. Its main features are:

- use of a strategy to limit the velocity of the particles,
- external archive to store the non-dominated solutions (leaders) and a density estimator (crowding distance),
- leader selection with a binary tournament from the leaders archive taking into account the crowding distance,
- mutation operator (polynomial mutation) that add turbulence.

It has been shown to exhibit high performances in various benchmarks and real-world problems [14], and suitable for large scale problems up-to 2048 variables [13]. Briefly as described in [14], the pseudo code of SMPSO is given in Algorithm 1.

Algorithm 1: SMPSO pseudo-code (from [14])

---

initializeSwarm()
initializeLeadersArchive()
generation = 0
while *generation < maxGenerations* **do**
  computeSpeedVector()
  updateParticlesPosition()
  mutation()
  evaluation()
  updateLeadersArchive()
  updateLocalBest()
  generation ++
end
returnLeadersArchive()

In the original publication, the performances of SMPSO were stated on benchmark problems after 25000 function evaluations [14]. Such a number of function evaluations seems to be not compatible with expensive problems. Motivated by decreasing the computational costs in evolutionary optimization of expensive problems, Surrogate-Assisted evolutionary computation has been highly investigated (see for review [9]). Basically, surrogates are used together with the real fitness function with the aim of estimating the fitness of new individuals of which can be then selected upon a given criteria for a real evaluation. As the real-world problem presented here is expensive, we modified SMPSO in order to add surrogates model (SASMPSO). The difference is that, during the evolutionary algorithm, the swarm particles will search according to their estimated evaluation thanks to surrogate, for a given amount of resources (*maxNfeSurrogate*). Once those resources are depleted, all the current swarm particles are evaluated using the real fitness function and then used in the evolutionary algorithm. Here, there is no particular selection of best candidates for real evaluations. The pseudo code of SASMPSO is given in Algorithm 2.

Algorithm 2: SASMPSO pseudo-code

initializeSwarm()
initializeLeadersArchive()
generation = 0
computeSpeedVector()
updateParticlesPosition()
mutation()
evaluation()
updateLeadersArchive()
generation ++
while *generation < maxGenerations* **do**
   selectSurrogates()
   trainSurrogates()
   nfeSurrogate = 0
   while *nfeSurrogate < maxNfeSurrogate* **do**
      computeSpeedVector()
      updateParticulesPosition()
      mutation()
      estimationWithSurrrogate()
      updateLocalBest()
      nfeSurrogate ++
   end
   evaluation()
   updateLocalBest()
   updateLeadersArchive()
   generation ++
end
returnLeadersArchive()

Upon the real evaluation of the initial swarm population, surrogates used here are constructed and updated for each objectives, separately, following the method described in [8]. Briefly, four models as Gaussian Process, Radial Basis Function, and Multivariate polynomial regression of degree 1 and 2 have been used here even if other surrogate models can be considered too. First, surrogates are selected based on the minimum root mean-squared error while performing a crossvalidation with 80% of samples for the training set and 20% of samples for the testing set (function *selectSurrogates* in Algorithm 2). Then, the selected surrogate is trained considering all the samples (function *trainSurrogates* in Algorithm 2). The samples are from an external archive different from the archive containing the leaders. This archive is updated with all the real-evaluated individuals and will grow until the algorithm stops. Whenever this archive is updated with new real-evaluated individuals, surrogates are re-selected and re-trained with the aim of getting the most accurate estimation.

4.1 Encoding/Decoding Parameters

For simplicity, parameters are encoded as float valued parameters in the range $[0, 1]$. In order to deal with the different ranges of the ISP parameters (see Sect. 5.1), a function is used to convert float values in $[0, 1]$ to proper parameters values compatible with the experimental ISP. To do so, Eq. 2 was used. Regarding the particular dependence between the parameters k and N of the denoising filter, Eq. 3 was used to compute the values of each k and N parameters before feeding the experimental ISP. One can note that N will not be tuned directly but k and n will be, thanks to Eq. 3.

$$x = (x_e * (max(x) - min(x))) + min(x), \text{ with } \begin{cases} x \in [\text{original range}] \\ x_e \in [0, 1] \end{cases} \quad (2)$$

$$N = 4 * n + k,$$
$$\text{with } \begin{cases} k = 2 * (round((k_e * (max(k) - min(k))) + min(k))) - 1, \, k_e \in [0, 1] \\ n = round((n_e * (max(n) - min(n))) + min(n)), \, n_e \in [0, 1] \text{ and } n \in [2, 5] \end{cases} \quad (3)$$

4.2 Function Evaluation

Algorithm 3: Function evaluation

$\underline{f}(rawImage, param_e)$;
Input: Raw image, encoded ISP parameters
Output: *acutance, noise*
$param = decode(param_e)$
$srbg = experimentalISP(rawImage, param)$
$acutance = (visualAcutance(srgb) - 102)^2$
$noise = visualNoise(srgb)$

The function evaluation is defined as in Algorithm 3. It takes as input the raw image ($rawImage$) and the ISP parameters. The encoded parameters ($param_e$) are decoded with the *decode* described function, into proper ranged values before feeding the experimental ISP. The output of the ISP, i.e. the processed image $srgb$ is then used to compute the image quality metrics with the functions *visualAcutance* and *visualNoise*. This gives the values of the two objectives that will be minimized by the EAs. One can note that the computation time of one function evaluation will be the sum of the ISP processing a raw image plus the metrics measure, that is 400 s in average.

5   Experimental Methods

Because of its complexity and as a first step, we propose to simplify the problem of ISP tuning by using EAs to optimize the parameters of a two-filtering stages ISP thanks to two image quality metrics. We designed a simplified two-filtering stages ISP that can be tuned according to 8 parameters, with respect to the minimization of 2 objectives. The parameters are related to the denoising and sharpening filters that will be detailed here after. As previously described, objectives that are minimized are the visual noise and the visual acutance. In the following we detail the experimental ISP, the implemented image filters and their parameters.

5.1   The Experimental ISP

As ISPs used in industry feature hundreds of parameters and are not open source, a simpler prototype has been developed. Based on commonly used procedures of raw image processing [16], we implemented a simple but fully functional ISP. Here as a proof of concept, only denoising and sharpening stages will be considered.

The denoise filter is based on the non-local means algorithm (NLM) proposed by [2]. Briefly, NLM filtering takes a mean of all pixels in the image, weighted by how similar these pixels are to the target pixel. This results in much greater post-filtering clarity, and less loss of detail in the image. Here is considered especially the implementation suggested in [4]. It depends on five parameters $\{\alpha, \beta, k, N, thr\}$ which are the lower/upper values $[\alpha, \beta]$ for the noise profile of the sensor itself that can vary in the 20% range, k and N the size of the patches of the denoise filter, and thr a threshold.

The sharpening stage is based on the commonly used unsharp-mask technique [7] where the original image is combined to a mask obtained from the negative image that is blurred, or "unsharp". This combination creates an output image that is less blurry than the original. This unsharp-mask depend on three parameters $\{radius, amount, threshold\}$ which are mainly related to the parameters of the Gaussian function used to blur the original image. All the eight parameters and their ranges are detailed in Table 1.

5.2   Raw Images

Nowadays, lots of imaging devices allow access to the raw output of the digital sensor such as digital single-lens reflex camera and even some recent smartphones with high-end photographic capabilities. Thanks to the suite [1], pictures of eCPIQeCPIQeCPIQtarget suitable for measuring the visual noise and the visual actuance (standard deadleaves target) have been made in a controlled environment. Especially, two raw images of this target, one in low-light condition (20 lux) and one in normal-light condition (100 lux) have been considered.

Table 1. Tunable parameters of the experimental ISP for denoising and sharpen filters.

Name	Range	Type
α	$\alpha \pm 20\%$	Float
β	$\beta \pm 20\%$	Float
k	$[3, 5, 7, 9]$	Integer
N	$N = f(k)$, with $N \leq 37$	Integer
thr	$[0, 2]$	Float
Radius	$[0, 3]$	Float
Amount	$[0, 2]$	Float
Threshold	$[0, 1]$	Float

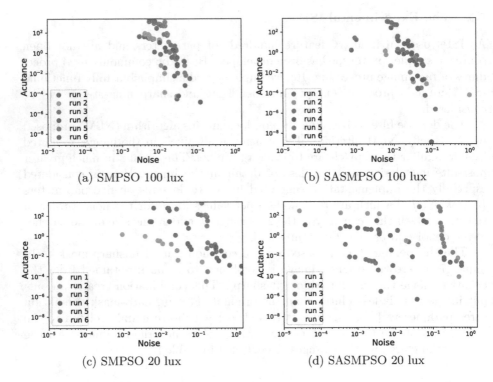

(a) SMPSO 100 lux   (b) SASMPSO 100 lux

(c) SMPSO 20 lux   (d) SASMPSO 20 lux

Fig. 2. Pareto fronts for each independent runs (one color per run), and for each algorithm and raw image combination, after 1000 nfe.

Typically, in low-light condition, the image is more noisy and should be more difficult to process than the normal-light condition image. Thus, the EA will optimize the parameters of the experimental ISP considering one of the two images at a time and results will be compared.

5.3   Performance Evaluation

In this study, both SMPSO and the proposed SASMPSO are compared for the optimization of the experimental ISP parameters considering two raw images (100 lux and 20 lux).

Hyperparameters are the same for both algorithms and raw images. The initial population size and the leaders archive size are set to 50. When surrogates are involved, the maximum number of evaluations using the surrogates (*maxNfeSurrogate*) is set to 1000. The maximum number of real evaluations allowed is set to 1000. Of course one could also vary those hyperparameters to analyze the sensitivity of the algorithms. But this is not the aim of the presented study. Six independent runs are performed. We agree that such a low number of runs is not enough to get statistically significant conclusions, and will need to be completed. But here are preliminary results that we think are already interesting for the community.

As the considered problem is black-box, the true Pareto front is unknown. Nevertheless, the performance between algorithms can be compared visually by the cumulative Pareto front across the 6 independent runs obtained after 1000 real evaluations, i.e. the non-dominated solutions over the concatenated solutions of the 6 runs. The idea beyond this is to have an overview of the exploration of the solution space.

The hypervolume indicator is computed to assess the differences in diversity and convergence. The hypervolume is the area between the non-dominated solutions and a reference point. Especially due to the low number of independent runs available, not the average but the median of the hypervolumes computed for the 6 runs is considered and is displayed as a function of the number of real function evaluations (nfe). The aim here is to provide a measure of central tendency: what is the most representative behavior of the hypervolume across the 6 runs. With such a low number of runs, the average will be highly sensitive to extreme values and variability unlike the median.

6   Results

Pareto fronts for each independent runs can be seen in Fig. 2. One can note the variabilty between the runs of one algorithm given one raw image, especially for SASMPSO 20 lux (Fig. 2d). Because of these discrepancies and the low number of available runs, not the mean but the median of the hypervolume is used.

For each combination algorithm-raw image, the hypervolume is computed for each run and the median is plotted as a function of the increasing number of nfe (Fig. 3a). For all, the hypervolume is increasing with the nfe showing that there is minimization of the two objectives. One can speculate that there is convergence toward regions of the solution space containing better solutions. Also here, little differences can be seen between the two algorithm while considering the 100 lux raw image. SASMPSO seems to perform better than SMPSO while considering the 20 lux raw image. But as a squared difference is used as the Acutance fitness (see 3), objective values evolve within a huge range. As a consequence, the

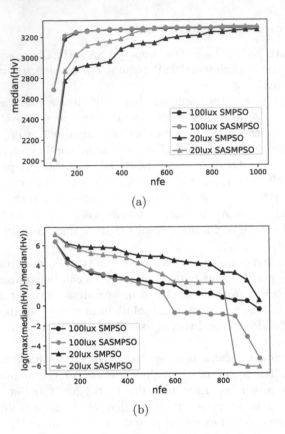

Fig. 3. (a) Median of the hypervolumes for the 6 independent runs. (b) *log* of the difference between the maximum of the median of hypervolumes and the median of hypervolumes. SMPSO is in black, SASMPSO is in grey, and for the raw images, 100 lux is circle markers and 20 lux is triangle markers.

reference point use to compute the hypervolume is far from the final solutions and thus the differences in hypervolumes that could be observed toward the "convergence" are smothered. In order to reduce this bias, is represented the *log* of the difference between the maximal median hypervolume and the considered hypervolume ($log(max(medians) - median)$) in Fig. 3b. As a result, it can be seen the decrease toward the highest hypervolume, i.e. the faster the decrease, the better. Thus, it is highlighted that SASMPSO (grey) is converging faster than SMPSO (black) toward this best encountered hypervolume. Also in both algorithms, the decrease toward the highest hypervolume seems to be slower while considering the 20 lux image compared to the 100 lux image (both circles versus both triangles curves until 800 nfe, Fig. 3b). This supports the *a priori* about the increased difficulty of the processing due to the inherent higher noise level in the 20 lux image.

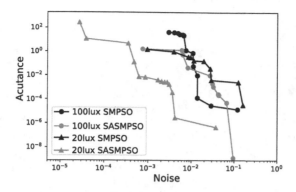

Fig. 4. Cumulative Pareto fronts across 6 independent runs after 1000 function evaluations for each algorithms and raw images combination. SMPSO is in black, SASMPSO is in grey. And for the raw images, 100 lux is circle markers and 20 lux is triangle markers.

Figure 4 shows the cumulative Pareto fronts across the 6 independent runs, after 1000 nfe for SMPSO (black), SASMPSO (grey), considering the 20 lux raw image (triangles) or the 100 lux raw image (circles). The objectives are in *log* scale. Both algorithms reach solutions in the range of expected values for both objectives. The use of surrogates in SASMPSO seems to lead to a better convergence for both raw images. Even if the differences are low for the 100 lux raw image, SASMPSO seems to outperform SMPSO for the enhancement of the 20 lux raw image, although the processing of the 20 lux raw image was *a priori* more difficult to optimize due to its inherent higher noise level compared to the 100 lux raw image. For illustration, crops of the raw image and the processed image related to the kneepoint solution of the cumulative Pareto front of each conditions are showed in Fig. 5.

7   Discussion and Conclusions

We presented in this study the use of an EA to optimize the parameters of a set of filters for the image enhancement, i.e. for the optimization of an ISP. Commonly in the literature, EAs were used in the context of image processing for the optimization or the design of single filters for image enhancement [3,5, 12,18]. To our knowledge, this is the first time that EAs are used to optimize a set of filters for image enhancement. Among the diversity of available EAs we focused on PSO, especially SMPSO [14] and the proposed simple surrogate-assisted SMPSO, SASMPSO, to optimize the parameters of an experimental ISP for the enhancement of raw images in normal-light and low-light conditions.

Despite the fact that the presented results can be considered as preliminary, results show that both EA succeeded in converging toward solutions in the expected range of values for the objectives, for both light conditions. Results show also that the use of surrogates allows to "converge" quicker. One could

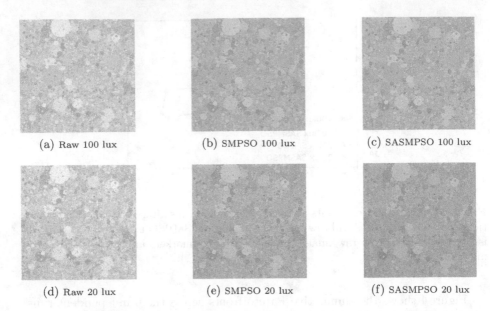

(a) Raw 100 lux   (b) SMPSO 100 lux   (c) SASMPSO 100 lux

(d) Raw 20 lux   (e) SMPSO 20 lux   (f) SASMPSO 20 lux

Fig. 5. Raw images and processed images related to the kneepoint solutions of the Pareto fronts showed in Fig. 4

argue that only a simplified ISP have been used here and actual complete ISP are far more complex and challenging. We agree that the presented results obtained using the simplified ISP need to be confirmed with complementary experiments using more complex ISP (more filters, more parameters, more image quality metrics). Indeed, besides the black-box and expensive aspects, complete ISPs can have approximately tens processing stages, each of them parameterized with tens or even hundreds of parameters. Moreover, one could consider that the sequence of processing stages itself, can be optimized which highly increase the complexity of the problem.

So, the final problem combines many of current challenges as black-box, expensive, large scale, many objectives optimization which have been and are currently investigated separately or partially combined, by the community. Going toward the optimization of a complete ISP will require the development of new surrogate-assisted EAs that can handle all those challenges at once.

References

1. IEEE CPIQ 1858–2016 - IEEE standard for camera phone image quality (2016). https://standards.ieee.org/standard/1858-2016.html
2. Buades, A., Coll, B.: A non-local algorithm for image denoising. In: CVPR, pp. 60–65 (2005)
3. Chen, J., Yu, W., Tian, J., Chen, L., Zhou, Z.: Image contrast enhancement using an artificial bee colony algorithm. Swarm Evol. Comput. **38**, 287–294 (2018)

4. Darbon, J., Cunha, A., Chan, T.F., Osher, S., Jensen, G.J.: Fast nonlocal filtering applied to electron cryomicroscopy. In: 2008 5th IEEE International Symposium on Biomedical Imaging: From Nano to Macro, pp. 1331–1334. IEEE (2008)
5. Ebner, M.: Engineering of computer vision algorithms using evolutionary algorithms. In: Blanc-Talon, J., Philips, W., Popescu, D., Scheunders, P. (eds.) ACIVS 2009. LNCS, vol. 5807, pp. 367–378. Springer, Heidelberg (2009). https://doi.org/10.1007/978-3-642-04697-1_34
6. Figueiredo, E., Ludermir, T., Bastos-Filho, C.: Many objective particle swarm optimization. Inf. Sci. **374**, 115–134 (2016). https://doi.org/10.1016/j.ins.2016.09.026, http://www.sciencedirect.com/science/article/pii/S0020025516308404
7. Gonzalez, R.C., Woods, R.E.: Digital Image Processing, 3rd edn. (2008)
8. Habib, A., Singh, H.K., Chugh, T., Ray, T., Miettinen, K.: A multiple surrogate assisted decomposition based evolutionary algorithm for expensive multi/many-objective optimization. IEEE Trans. Evol. Comput. **23**, 1000–1014 (2019)
9. Jin, Y.: Surrogate-assisted evolutionary computation: recent advances and future challenges. Swarm Evol. Comput. **1**(2), 61–70 (2011)
10. Karaboga, D.: An idea based on honey bee swarm for numerical optimization. Technical report (2005)
11. Kennedy, J., Eberhart, R.: Particle swarm optimization (PSO). In: Proceedings of IEEE International Conference on Neural Networks, Perth, Australia, pp. 1942–1948 (1995)
12. Munteanu, C., Rosa, A.: Towards automatic image enhancement using genetic algorithms. In: Proceedings of the 2000 Congress on Evolutionary Computation, CEC 2000 (Cat. No. 00TH8512), vol. 2, pp. 1535–1542. IEEE (2000)
13. Nebro, A.J., et al.: Extending the speed-constrained multi-objective PSO (SMPSO) with reference point based preference articulation. In: Auger, A., Fonseca, C.M., Lourenço, N., Machado, P., Paquete, L., Whitley, D. (eds.) PPSN 2018. LNCS, vol. 11101, pp. 298–310. Springer, Cham (2018). https://doi.org/10.1007/978-3-319-99253-2_24
14. Nebro, A.J., Durillo, J.J., Garcia-Nieto, J., Coello, C.C., Luna, F., Alba, E.: SMPSO: a new PSO-based metaheuristic for multi-objective optimization. In: 2009 IEEE Symposium on Computational Intelligence in Multi-Criteria Decision-Making (MCDM), pp. 66–73. IEEE (2009)
15. Reyes-Sierra, M., Coello, C.C., et al.: Multi-objective particle swarm optimizers: a survey of the state-of-the-art. Int. J. Comput. Intell. Res. **2**(3), 287–308 (2006)
16. Rob, S.: Processing raw images in MATLAB (2014). http://rcsumner.net/raw_guide/RAWguide.pdf
17. Sahoo, M.: Classical and evolutionary image contrast enhancement techniques: comparison by case studies. In: Behera, H.S., Mohapatra, D.P. (eds.) Computational Intelligence in Data Mining. AISC, vol. 556, pp. 37–44. Springer, Singapore (2017). https://doi.org/10.1007/978-981-10-3874-7_4
18. Yang, X.S., Papa, J.P.: Bio-Inspired Computation and Applications in Image Processing. Academic Press, Cambridge (2016)
19. Zhang, Y., Gong, D.W., Geng, N.: Multi-objective optimization problems using cooperative evolvement particle swarm optimizer. J. Comput. Theor. Nanosci. **10**(3), 655–663 (2013)

Combinatorial Surrogate-Assisted Optimization for Bus Stops Spacing Problem

Florian Leprêtre[✉], Cyril Fonlupt, Sébastien Verel, and Virginie Marion

Univ. Littoral Côte d'Opale, LISIC, 62100 Calais, France
florian.lepretre@univ-littoral.fr

Abstract. The distribution of transit stations constitutes an ubiquitous task in large urban areas. In particular, bus stops spacing is a crucial factor that directly affects transit ridership travel time. Hence, planners often rely on traffic surveys and virtual simulations of urban journeys to design sustainable public transport routes. However, the combinatorial structure of the search space in addition to the time-consuming and black-box traffic simulations require computationally expensive efforts. This imposes serious constraints on the number of potential configurations to be explored. Recently, powerful techniques from discrete optimization and machine learning showed convincing to overcome these limitations. In this preliminary work, we build combinatorial surrogate models to approximate the costly traffic simulations. These so-trained surrogates are embedded in an optimization framework. More specifically, this article is the first to make use of a fresh surrogate-assisted optimization algorithm based on the mathematical foundations of discrete Walsh functions in order to solve the real-world bus stops spacing optimization problem. We conduct our experiments with the SIALAC benchmark in the city of Calais, France. We compare state-of-the-art approaches and we highlight the accuracy and the optimization efficiency of the proposed methods.

Keywords: Bus stops spacing · Combinatorial optimization · Surrogate models

1 Motivations

The United Nations expect sixty percent of the world's population to live in urban areas by the next decade [2]. This relentlessly growing rate constantly challenges urban planners to design sustainable cities so as to improve the mobility of their inhabitants and travellers. This objective can be achieved, in a way, by an efficient planning and management of public transport systems, such as trams, buses or even self-service bicycles. The correct design of these systems is the key to offer potential users a competitive transit mode compared to the private car. This is particularly advocated for mitigating environmental impacts of transport and could also help to revitalize and renew interest in some districts

© Springer Nature Switzerland AG 2020
L. Idoumghar et al. (Eds.): EA 2019, LNCS 12052, pp. 42–52, 2020.
https://doi.org/10.1007/978-3-030-45715-0_4

of the city. For example, the location of a transit station in an area where none currently exists may attract new customers to public transport who previously lacked the service.

Substantial works exist in the litterature on the deployment of such efficient transport systems. A recurring challenge is to ensure that transit stops are properly spaced. Without loss of generality, the present article focuses on the optimization of bus stops spacing. It is well known that their distribution represents an important factor directly affecting passengers travel times [21,27]. Some research even investigates social costs, economic benefits or environmental impacts of bus stations positionings [19,24]. Also, several methods are introduced to find optimal spacings for minimal travel costs. On the one hand, coverage models have been widely employed, such as the Thiesen polygons [27] or the Voronoi diagrams [28]. These techniques allow additional data to be assessed while searching for optimal stop positions (e.g., district density around the stop). On the other hand, the discretization of the study areas has also shown promising. Ibeas et al. proposed to split the transit route area in small links of equal distance [11]. Each link represents a potential stop location, and the so-discretized optimization problem is solved with a pattern-search iterative algorithm [9]. Besides, Furth et al. considered each intersection of the studied road network as a potential stop location [8]. Then, a dynamic programming algorithm was used to determine the optimal bus stop positions.

Inspired by the aforementioned works, this article considers the bus stops spacing problem as a pseudo-boolean problem, where the passenger travel time is the fitness function to minimize. Then, a binary variable is associated with each possible location for a bus stop: the variable equals one if the stop is activated or equals zero if not – this will be discussed further in Sect. 3. However, such studies as well as many others in the literature mostly rely on numerical simulations of urban traffic flows, which are usually blackbox models. As a consequence, only the design variables and the resulting values of the simulation are known. Moreover, it is often computationally time expensive (from minutes to hours) to get the fitness value of one single simulation [3,6]. In addition with the combinatorial explosion of the search space, optimization experts thus face a serious limitation on the capacity to freely explore potential solutions. To tackle such an optimization challenge, one classical solution in Surrogate-assisted Optimization (SaO) is to learn a surrogate model to approximate the costly simulator evaluations and then reduce the number of potential sampled solutions during the search process. Although the field of combinatorial surrogate models has long received little attention, it is now experiencing a sudden renewed interest, bringing with it new algorithmic ideas to the community [4,5,22].

As part of this preparatory and applicative work for further studies, we implement a variety of recent combinatorial surrogate modeling techniques to approximate the time-expensive traffic simulations. Further, we embed the so-learnt surrogates in the context of bus stops spacing optimization. In particular, this work is the first to make use of a newly published SaO algorithm based on the mathematical foundations of discrete Walsh functions coupled with powerful grey-box optimization techniques [14], in order to solve a class of real-world problems. We aim to highlight the accuracy and the optimization performances of these methods.

The rest of this paper is organized as follows. In Sect. 2, we introduce the state-of-the-art combinatorial surrogate modeling techniques and the foundations of SaO. Sect. 3 is devoted to experiments specifications and their analyses. In Sect. 4, we conclude the paper and discuss future works.

2   Combinatorial Surrogate-Assisted Optimization

Surrogate models formulate quick-to-evaluate mathematical models, so as to approximate black-box and time-consuming computations. They are built from a sample of evaluated solutions. Therefore, the main purpose of surrogate-assisted optimization is to efficiently select the solutions to be sampled in order to quickly improve the quality of the surrogate and thus the quality of the solutions found.

A surrogate-assisted method combines three components (see Algorithm 1). The first component is the surrogate model itself which is a regression model of the fitness function. The model must be expressive enough to catch the complexity of the fitness function, but at the same time slightly sophisticated in order to ease the learning when a small sample of solutions is available. The second component is an acquisition function defined from the surrogate model. This acquisition function can be directly the surrogate model or a trade-off between the predicted quality of candidate solution and the estimation error of the surrogate model. The goal is to guide the search and to ensure a balance between exploration that increases the quality of the surrogate model and exploitation that pushes towards high-quality solutions according to the surrogate model. The last component is the algorithm to optimize the acquisition function. This algorithm has to be efficient in time and in quality to converge quickly to the promising solutions given by the acquisition function. Therefore, such promising solution is selected, evaluated and added to the sample for the next iteration of the search algorithm. An efficient surrogate-assisted optimizer for combinatorial problems is a relevant combination of these three components.

To the best of our knowledge, four main methods have been proposed for pseudo-boolean problems: Radial Basis Function model [16], Kriging approach [26], Bayesian approach [4] and Walsh basis functions decomposition [22].

Algorithm 1: Surrogate-assisted optimization framework.

---

1   $S \leftarrow$ Initial sample $\{(x, f(x)), \ldots\}$
2   **while** *computational budget is not spent* **do**
3 | $M \leftarrow$ Build model S
4 | $x \leftarrow$ Optimize M w.r.t. an acquisition function
5 | Evaluate x using f
6 | $S \leftarrow S \cup \{(x, f(x))\}$
7   **end**

Kriging Approach. Kriging approach is a direct extension of the numerical surrogate approach and is based on Gaussian Process (GP) [25,26]. In the context of combinatorial structures, Euclidean distance is replaced by the Hamming distance for pseudo-boolean functions or more sophisticated discrete distance for the search space of permutations. Then, Kriging is coupled with the Efficient Global Optimizer framework (EGO) [12]. This framework takes advantage of the uncertainty of the approximations given by the GP and uses a genetic algorithm to select the promising solution that maximizes the Expected Improvement (EI), *i.e.*, the acquisition function. One should note that the computational complexity of EI is high and can not be reduced by some classic techniques in combinatorial optimization such as incremental evaluation. Nevertheless, this surrogate-assisted optimization has been shown to outperform the aforementioned Radial Basis Function model [16] – thus the latter will not be detailed here.

Bayesian Approach. Another state-of-the-art surrogate-assisted approach is the Bayesian Optimization of Combinatorial Structures (BOCS) algorithm [4]. The statistical model of BOCS is the standard multilinear polynomial of binary variables. Therefore, Baptista *et al.* argue that the model takes into account the interactions between the binary variables. Only a quadratic polynomial model is studied in their article (*i.e.*, one variable interacts with only other one):

$$\forall x \in \{0,1\}^n, \ M_2(x) = a_0 + \sum_{i \in N} a_i x_i + \sum_{i < j \in N} a_{ij} \ x_i x_j, \tag{1}$$

where $N = \{1, \ldots, n\}$. The regression technique is the Sparse Bayesian Linear Regression [15]. The optimizer is a basic simulated annealing that minimizes the approximation of the fitness function provided by the surrogate model with a regularization term.

Walsh Basis Functions. A new combinatorial surrogate model based on Walsh functions has been proposed [22]. Walsh functions [23] describe a normal and orthogonal basis of discontinuous functions that can be employed to decompose any function of the Hilbert space. Therefore, Verel *et al.* assumed that the expensive pseudo-boolean functions might be substituted by a polynomial of Walsh decompositions of order k:

$$\forall x \in \{0,1\}^n, \ W_k(x) = \sum_{\ell \ s.t. \ o(\ell) \leqslant k} w_\ell \cdot (-1)^{\sum_{i=1}^n \ell_i x_i}, \tag{2}$$

where o is the order of the Walsh function, *i.e.*, the number of binary digits equals to 1 in the binary representation of ℓ. In the following, we restrain to quadratic interactions:

$$\forall x \in \{0,1\}^n, \ W_2(x) = w_0 + \sum_{i=1}^n w_i \ (-1)^{x_i} + \sum_{i < j \in N} w_{ij} \ (-1)^{x_i + x_j}. \tag{3}$$

To face the quadratic number of polynomial terms, the regression technique is a linear model trained with ℓ_1-norm as regularizer, aka the Lasso [20]. Recently, the Walsh Surrogate-assisted Optimization (WSaO) algorithm has been introduced [14]. The authors benefit from powerful grey-box optimization techniques and use the so-called Efficient Hill-climber (EH) from Chicano *et al.* [7] as an optimizer for the Walsh surrogates. WSaO has been shown to outperform both Kriging and BOCS approaches, scaling up at least to dimension $n = 100$.

3   Experiments

3.1   Overview

We consider the bus stops spacing problem as a pseudo-boolean optimization challenge. For the sake of simplicity in this preliminary work, the focus is on one regular bus line in the city of Calais, France.

Bus Stops. All potential bus stops are implemented in advance on the given road network. They are manually located on intersection nodes and are numbered from 1 to n, where n is the total number of potential stops in the bus route, following [8]. Then, a solution to the optimization problem, *i.e.*, a possible design of the bus stops, is denoted by a binary string $x \in \{0,1\}^n$. Therefore, open bus stops are associated to bits in x equal to one, whereas closed stops are associated to bits equal to zero. The first and last stops are constrained to be open. Only open bus stops are taken into account during the forthcoming simulations. Figure 1 illustrates a simplified bus stops design on the studied bus route. The complete bus route considers $n = 20$ potential stop locations.

Urban Flows. The simulation system considered in this work is the Multi Agent Transport Simulation (MATSim) [10]. MATSim requires as inputs a road network model [18] and the initial mobility scenarios for a set of agents (*i.e.*, a set of travelers' schedules). These scenarios are generated according to the SIALAC Benchmark [13]. The latter allows to synthesize mobility plans which assess such information as living quarters, business districts or main entry and exit points of the city. As an example, Fig. 2 illustrates a scenario where agents are distributed into four living quarters. The present work studies six scenarios involving 5000 travelers with different number of home and working area clusters. The possible home cluster number is 1 cluster, 4 clusters or uniform when the population is randomly distributed over the city area; they are denoted as $1h$, $4h$ or uh, respectively. The possible working area cluster number is 1 or 4 clusters; they are denoted as $1a$ or $4a$, respectively. Finally, a scenario defines a round trip between a home location and an activity location, for each traveler. The latter are distributed according to the configuration of the six studied scenarios: $1h$-$1a$, $1h$-$4a$, $4h$-$1a$, $4h$-$4a$, uh-$1a$, and uh-$4a$. The number of cluster impacts the travel time for pedestrians and cars which allows to test the robustness of SaO algorithms. One simulation with MATSim, *i.e.*, one fitness function evaluation, requires about a minute of computation as a single-thread program.

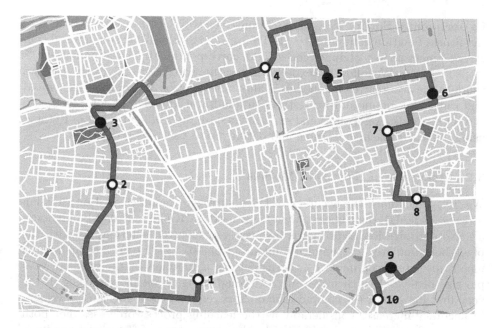

Fig. 1. Simplified example of open or closed bus stops (white or black dots) on the regular bus route (red line), according to the solution $x = 1101001101$. (Color figure online)

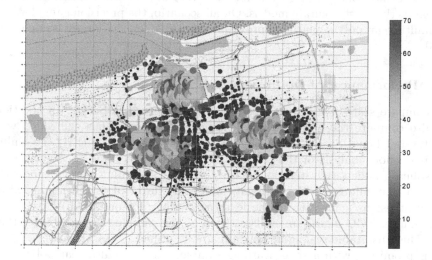

Fig. 2. Four home clusters inside Calais road network. Colors indicate the number of agents departing from a node. (Color figure online)

Experimental Setup. We aim to minimize the travelers mean travel time, *i.e.*, the fitness function computed by MATSim, *i.e.*, the black-box simulator. Then, we use binary strings of dimension $n = 20$, corresponding to the 20 potential bus stops locations. We follow the experimental setup exposed in [14], except that we validate the accuracy of the models against a test-set of 250 solutions generated uniformly at random and we restrain to quadratic interactions for BOCS and WSaO methods. Algorithms and experiments are fully implemented in Python, using standard machine learning and optimization packages [17].

3.2   Accuracy of Surrogate Models

We first benchmark the accuracy of three surrogate models based on gaussian process (Kriging), multilinear polynomials used in BOCS, and Walsh polynomials. No optimization algorithm is involved yet. Selected solutions to learn the black-box simulator are sampled randomly from $\{0, 1\}^n$. Figure 3 compares the mean absolute error made by the models as a function of the random sample size dedicated to their learning. Although Kriging seems promising in the very first iterations, the computational effort required for a slight improvement in accuracy increases considerably as the learning process progesses. For both Walsh, and multilinear based methods, the convergence of the model quality is reached around a sample size of 400 solutions on 3 scenarios (1h-1a, 4h-1a and uh-1a). For more difficult scenarios 1h-4a, 4h-4a and uh-4a, the precision quality increases beyond the largest sample size. Overall scenarios, Walsh surrogates appear as the most accurate models. On all scenario, the precision gain of Walsh *vs.* multilinear polynomial is approximatively 30% for the largest sample size of 10^3.

3.3   Performances of Optimizers

We compare state-of-the-art SaO algorithms presented in Sect. 2. In addition, we also compare the performances of multilinear polynomials embedded with an Iterated Local Search based on the Efficient Hill-climber (EH) [7]. Unlike the regression analysis of the previous section and according to Algorithm 1, the solution added each iteration to the surrogate's learning sample is now the solution that minimizes the surrogate model. Figure 4 plots the minimization of the fitness as a function of the learning sample, for one SIALAC scenario. At a glance, EGO algorithm stalls as soon as the sample size gets bigger than 100. This unsuitable scaling-up was already identified in [14]. However, WSaO seems promising with a small learning sample, while a multilinear polynomial coupled with BOCS appears better when the sample size grows (see Table 1). However, notice that the order of difference between the two approaches is only in seconds of mean travel time. Both methods seems promising for that moderate size scenario with one bus line, and 20 potential bus stop positions. These results are still under active work and follow the first results performed on artificial benchmarks [14].

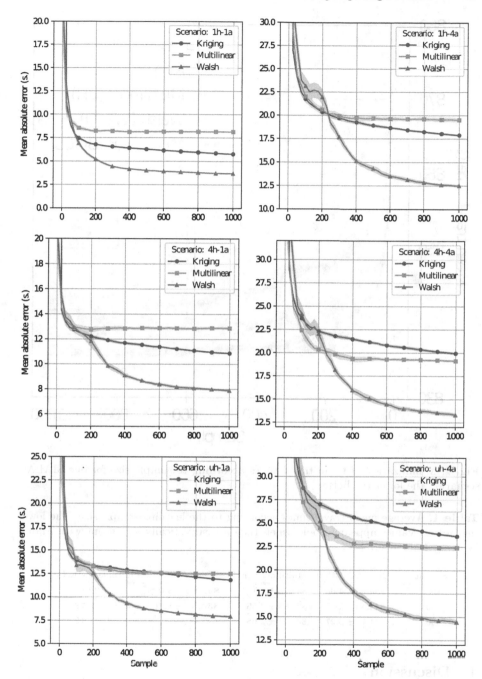

Fig. 3. Mean absolute error with confidence interval as a function of random samples for six SIALAC scenarios (h: home clusters, a: activity clusters). The lower the better.

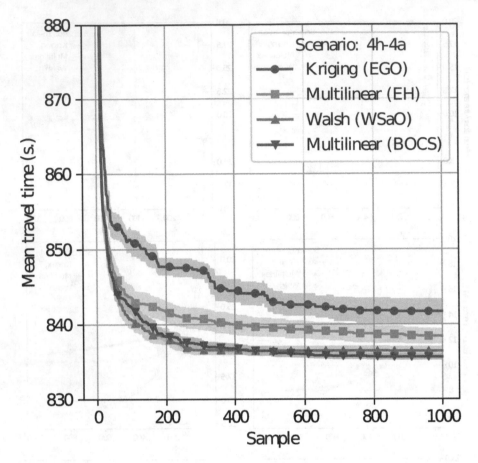

Fig. 4. Minimization of mean travel times according to sample size for one SIALAC scenario. The lower the better.

Table 1. Average mean travel times according to the learning sample size. The lower the better. Values appear in bold when they are statistically significant with Mann-Whitney tests at level 5%.

Sample size	Kriging (EGO)	Multilinear (EH)	Walsh (WSaO)	Multilinear (BOCS)
100	850.82 ± 4.00	843.55 ± 4.17	**840.59 ± 3.03**	842.24 ± 2.58
400	844.49 ± 3.55	839.99 ± 3.56	**836.57 ± 1.70**	**836.81 ± 1.78**
1000	841.71 ± 3.27	838.41 ± 2.68	836.43 ± 1.62	**835.64 ± 1.13**

4 Discussion

Combinatorial surrogate models succeed to learn the time-consuming and black-box traffic simulator, with a reasonable error lower than three percent of the average real simulator responses. The results show that polynomial-based models coupled with grey-box optimization algorithms are competitive against standard state-of-the-art methods in a surrogate-assisted optimization purpose. EGO approach quickly stalls, while WSaO and BOCS approaches converge to satisfying bus route designs. In particular, this article is the first to apply WSaO to solve such a class of real-world optimization problem.

This preparatory work opens many directions for future applicative research. First, we would like to scale up the bus stop spacing problem to a hundred dimensions at least, in order to design more concise bus routes and to get more challenging problems to face for state-of-the-art SaO methods. As it was pointed out in [14], a higher dimension would allow to draw more clearly conclusions as to the most appropriate polynomial decomposition. Further, these polynomial models are restrained here to quadratic interactions between variables. The cubic, or higher interaction is envisaged in order to aspire to the conception of more accurate models. Finally in a more applicative way, we are considering to redraw a part of the bus routes, based on the results obtained from SaO algorithms. Such routes could be implemented in the Zenbus [1] vizualisation platform, which could represent a powerful tool for urban planners and decision-makers.

Acknowledgments. Experiments presented in this paper were carried out using the CALCULCO computing platform, supported by SCOSI/ULCO (Service COmmun du Système d'Information de l'Université du Littoral Côte d'Opale). We are grateful to PMCO for its funding, and we thank Calais city (France) for the data and its support.

References

1. Zenbus. https://zenbus.net/sitac-calais. Accessed 31 May 2019
2. The world's cities in 2016. United Nations, Department of Economic and Social Affairs (2016)
3. Armas, R., Aguirre, H., Zapotecas-Martínez, S., Tanaka, K.: Traffic signal optimization: minimizing travel time and fuel consumption. In: Bonnevay, S., Legrand, P., Monmarché, N., Lutton, E., Schoenauer, M. (eds.) EA 2015. LNCS, vol. 9554, pp. 29–43. Springer, Cham (2016). https://doi.org/10.1007/978-3-319-31471-6_3
4. Baptista, R., Poloczek, M.: Bayesian optimization of combinatorial structures. In: International Conference on Machine Learning (ICML), pp. 462–471 (2018)
5. Bartz-Beielstein, T., Zaefferer, M.: Model-based methods for continuous and discrete global optimization. Appl. Soft Comput. **55**, 154–167 (2017)
6. Branke, J.: Simulation optimization tutorial. In: Proceedings of the Genetic and Evolutionary Computation Conference Compagnion. ACM (2018)
7. Chicano, F., Whitley, D., Sutton, A.M.: Efficient identification of improving moves in a ball for pseudo-Boolean problems. In: Proceedings of the 2014 Annual Conference on Genetic and Evolutionary Computation, GECCO 2014, pp. 437–444. ACM, New York (2014)
8. Furth, P., Rahbee, A.B.: Optimal bus stop spacing through dynamic programming and geographic modeling. Transp. Res. Rec. **1731**, 15–22 (2000)
9. Hooke, R., Jeeves, T.A.: "Direct search" solution of numerical and statistical problems. J. ACM **8**(2), 212–229 (1961)
10. Horni, A., Nagel, K., Axhausen, K. (eds.): Multi-Agent Transport Simulation MATSim. Ubiquity Press, London, August 2016

11. Ibeas, Á., dell'Olio, L., Alonso, B., Sainz, O.: Optimizing bus stop spacing in urban areas. Transp. Res. Part E: Logistics Transp. Rev. **46**(3), 446–458 (2010)
12. Jones, D.R., Schonlau, M., Welch, W.J.: Efficient global optimization of expensive black-box functions. J. Glob. Optim. **13**(4), 455–492 (1998)
13. Leprêtre, F., Fonlupt, C., Verel, S., Marion, V.: SIALAC benchmark: on the design of adaptive algorithms for traffic lights problems. In: Proceedings of the Genetic and Evolutionary Computation Conference Companion, pp. 288–289. ACM (2018)
14. Leprêtre, F., Fonlupt, C., Verel, S., Marion, V.: Walsh functions as surrogate model for pseudo-Boolean optimization problems. In: Proceedings of the Genetic and Evolutionary Computation Conference. ACM (2019)
15. Makalic, E., Schmidt, D.F.: A simple sampler for the horseshoe estimator. IEEE Sign. Process. Lett. **23**(1), 179–182 (2016)
16. Moraglio, A., Kattan, A.: Geometric generalisation of surrogate model based optimisation to combinatorial spaces. In: Merz, P., Hao, J.-K. (eds.) EvoCOP 2011. LNCS, vol. 6622, pp. 142–154. Springer, Heidelberg (2011). https://doi.org/10.1007/978-3-642-20364-0_13
17. Pedregosa, F., et al.: Scikit-learn: machine learning in Python. J. Mach. Learn. Res. **12**, 2825–2830 (2011)
18. Ramm, F., Karch, C., Topf, J.: Geofabrik. https://www.geofabrik.de. Accessed 24 Jan 2018
19. Saka, A.A.: Model for determining optimum bus-stop spacingin urban areas. J. Transp. Eng. **127**(3), 195–199 (2001)
20. Tibshirani, R., Wainwright, M., Hastie, T.: Statistical Learning with Sparsity: The Lasso And Generalizations. Chapman and Hall/CRC, London (2015)
21. Vaughan, R., Cousins, E.: Optimum location of stops on a bus route. In: 1977 7th International Symposium on Transportation and Traffic Theory, Kyoto, Japan (1977)
22. Verel, S., Derbel, B., Liefooghe, A., Aguirre, H., Tanaka, K.: A surrogate model based on walsh decomposition for pseudo-boolean functions. In: Auger, A., Fonseca, C.M., Lourenço, N., Machado, P., Paquete, L., Whitley, D. (eds.) PPSN 2018. LNCS, vol. 11102, pp. 181–193. Springer, Cham (2018). https://doi.org/10.1007/978-3-319-99259-4_15
23. Walsh, J.L.: A closed set of normal orthogonal functions. Am. J. Math. **45**(1), 5–24 (1923)
24. Wirasinghe, S.C., Ghoneim, N.S.: Spacing of bus-stops for many to many travel demand. Transp. Sci. **15**(3), 210–221 (1981)
25. Zaefferer, M., Stork, J., Bartz-Beielstein, T.: Distance measures for permutations in combinatorial efficient global optimization. In: Bartz-Beielstein, T., Branke, J., Filipič, B., Smith, J. (eds.) PPSN 2014. LNCS, vol. 8672, pp. 373–383. Springer, Cham (2014). https://doi.org/10.1007/978-3-319-10762-2_37
26. Zaefferer, M., Stork, J., Friese, M., Fischbach, A., Naujoks, B., Bartz-Beielstein, T.: Efficient global optimization for combinatorial problems. In: GECCO (2014)
27. Zheng, C., Zheng, S., Ma, G.: The bus station spacing optimization based on game theory. Adv. Mech. Eng. **7**(2), 453979 (2015)
28. Zhu, Z., Guo, X., Chen, H., Zeng, J., Wu, J.: Optimization of urban mini-bus stop spacing: a case study of Shanghai (China). Tehnicki Vjesnik **24**, 949–955 (2017)

Optimisation of a Checkers Player Using Neural and Metaheuristic Approaches

Ethan Bunce[(✉)] and Edward Keedwell

College of Engineering, Mathematics and Physical Sciences,
University Exeter, Exeter EX4 4QF, UK
ethan.bunce@hotmail.com

Abstract. Within this paper we evaluate the components used to build a checkers playing system with no embedded expert knowledge. We found that Particle Swarm Optimisation (PSO) and Evolutionary Algorithms (EA) are suitable training methods for Artificial Neural Networks (ANN) acting as evaluation functions within minimax, when training on 2 and 4 plies. By playing the trained networks against one other the single best network was found, which was produced by 3000 iterations of PSO playing on 2 plies. We show that this network outperformed a piece differential evaluation function, both on a fixed number of plies, and when using iterative deepening search. We also show that the higher the amount of plies the better a system will perform, however it is the relative difference between the amount of plies that impacts the performance. External validation of the system shows it winning all 44 games it played against non-expert human players. It was also able to solve the hardest tasks on a checkers problem website. The system was also able to draw against Chinook, a checkers playing system with expert knowledge and state-of-the-art in the field.

Keywords: Evolutionary Algorithm · Particle Swarm Optimisation · Artificial Neural Network · Game theory

1 Introduction

This paper investigates metaheuristic and neural networks methods behind developing a game playing system with no expert knowledge, and apply them to the board game checkers. A similar attempt was undertaken by Kumar Chellapilla and David Fogel in their 1999 paper *"Evolving neural networks to play checkers without relying on expert knowledge"* [6]. As with Chellapilla and Fogel's paper we used an Artificial Neural Network (ANN) as an evaluation function within Minimax to play the game. Minimax enables a machine to play turn-based games by assuming perfect play from the opponent and then through tree search, proposing the move that minimises the possible loss in the worst case scenario. The method has been shown to be effective, but is highly dependent on

© Springer Nature Switzerland AG 2020
L. Idoumghar et al. (Eds.): EA 2019, LNCS 12052, pp. 53–67, 2020.
https://doi.org/10.1007/978-3-030-45715-0_5

the utility function which evaluates the game state once move has been made. In this paper we extend [6] by implementing different methods for training ANNs as the utility function, as well as other elements of game theory to increase the performance of the system.

The gameplaying method used here is to allow players to play games against each other and to learn from the wins and losses that result, a method that requires no expert knowledge of the rules to be used. For games we already understand, it can enable us to learn new strategies and methods, which can be far superior to existing human ones. A recent example of this is AlphaGo Zero, a Go game playing system developed by Google's Deep Mind, which in 40 days was able to overtake all human players and reach an ELO rating of 5,185 [15], far higher than the current human highest rating of 3,646 held by South Korean player Shin Jinseo [2].

Within this paper we have investigated, and compared the performance of, EA and PSO for training ANNs to be used as evaluation functions within minimax for the board game checkers [4,7,12]. We have also investigated the impact on system performance of the number of moves ahead (known as the ply) searched by the algorithm.

2   Background

Checkers. The game used here is American checkers, also known as British draughts, which is played on an eight by eight board. The rules can be viewed in full at [3].

Previous Game Playing Systems. Deep Blue, a Chess game playing system, developed by IBM which in 1997 became the first system to beat a world champion at chess. It used Alpha-Beta pruning search on a minimax tree, running the search in parallel, with an evaluation function designed with expert knowledge [5,8]. In addition to this it used an endgame database (EGDB). The EGDB contains all possible board positions with less than a certain number of pieces remaining, alongside whether it is a win loss or draw. This allows the system to play perfectly, only selecting moves that will maintain its winning position.

Chinook is a checkers game play system, developed over 18 years at the University of Alberta. The evaluation function uses expert knowledge such as trapped kings and runaway checkers (a man who has an unimpeded path to the opposing players kings row). In 1994 it won the World Championship of Checkers, being the first computer to do so. After this the project retired from playing humans to instead solve checkers. At the time of its retirement it was rated at 2814 ELO. Since then Chinook and its authors have solved checkers [14], proving that it is unbeatable with perfect play, and the best result an opposing player can achieve is a draw. Chinook used many similar features to Deep Blue, including an opening book, endgame database, an expertly crafted evaluation function and Alpha-Beta pruning.

Training Methods. Due to the nature of the project it would not be possible to use traditional training methods. For example it is common to use backpropagation to train a neural network, however this requires known input/output pairs which are not available in the adversarial training method described above. Instead we focused on two training methods, evolutionary algorithms (EA) and particle swarm optimisation (PSO) [9,16] to train the weights of the ANN utility function.

Game Theory. In order to further increase the performance of the system we included other elements of game theory. From looking at previous game playing systems it was clear that it was standard to use Alpha-Beta pruning [10], iterative deepening search [11] and book moves. Alpha-Beta pruning decreases execution time, without affecting the results of the search. Iterative deepening allows the system to take full advantage of all the time it has per move, whilst ensuring the current best move is returned within the time limit. Book moves allow for perfect play at specific parts of the game.

3   Methodology

The system used a trained ANN as a utility function within minimax. The ANN consisted of 91 input nodes, 2 hidden layers with a size of 40 and 10 nodes and single output node. The input consisted of the piece differential of each subsquare on the board, ranging in size from 3×3 up to the full board. The full board piece differential was also passed as an input to the output node.

3.1   Minimax

The system used minimax to choose which moves to make. It can search to a fixed depth, or as seen in some of the experiments use an iterative deepening search, in order to make full use of its time per move. Using iterative deepening also allows us to look at the impact on execution time, and thus performance, from using a neural network as a utility function.

3.2   Artificial Neural Networks

Both training methods trained networks of a fixed topology. The network architecture is similar to Chellapilla and Fogel's architecture [6]. From reading their paper it was unclear if each subsection connected to one input node, or each subsection of the same size connects to all the input nodes for that size (i.e. 36 nodes for 3×3, 25 nodes for 4×4 etc.). We initially tested with the former, however found the bias had too much of an impact on the output, and changed to using the latter. The network consists of the input nodes described (91), which feed into 40 hidden nodes, followed by another 10 hidden nodes and finally a single output node. When calculating the inputs a higher value can be placed on the king, as it is a more advantageous piece to possess. This value (known as the king value) was modifiable by the training methods. The output node is also supplied with the total piece differential of the entire board. Each node used *tanh* as the activation function (bounded between −1 and 1). The value outputted by this node is the value that is then used within minimax. The diagram can be seen in Fig. 1.

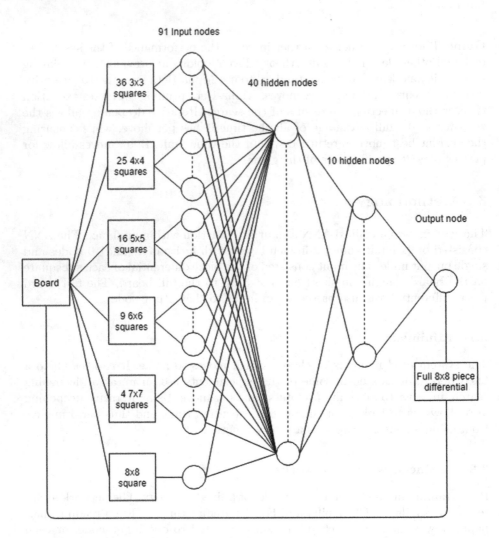

Fig. 1. The network architecture used within the paper

3.3   Training Methods

Two training methods were implemented, particle swarm optimisation (PSO) and evolutionary algorithms (EA). Both training methods awarded 1 point for a win, -2 for a loss, and a value between 1 and -2 for a draw, depending on the final piece differential, with a more favourable piece differential being awarded a higher score.

The network was encoded as an object in the program, with an array of neurons forming a layer, and an array of layers forming a full network. Each of the neurons had an array of decimal values for weights, and a single decimal value as a bias. The training methods accessed these instance variables in order to produce new solutions.

EA. The evolutionary algorithm is similar to that used by Chellapilla and Fogel in their paper [6]. The algorithm began with a randomly created population of 15 networks. The weights and biases for these networks were sampled uniformly over $[-0.2, 0.2]$, in order to provide a small amount of variance between them, and the king value was set to 2. The self-adaptive parameter σ for each weight and bias was initialized to 0.05. Each parent in the population produced new networks by varying the weights, biases and king value. The new network n^+ was created by:

$$\sigma^+(j) = \sigma(j)exp(\tau N_j(0,1)), \quad j = 1, ..., N_w \qquad (1)$$

$$w^+(j) = w(j) + \sigma^+(j)N_j(0,1), \quad j = 1, ..., N_w \qquad (2)$$

where N_w is the total number of weights and biases in the network, 6468, $\tau = 1/\sqrt{(2 \times \sqrt{(N_w)})} = 0.07885$, and $N_j(0,1)$ is a standard Gaussian random variable resampled for each j. The king value K^+ is obtained by taking the original king value K, and adding a number chosen randomly from -0.1, 0, 0.1. The king value was constrained to between 1 and 3. Each network in the population would play 5 games against random opponents from the population, playing as white. After this the 15 networks with the lowest scores were removed from the population and new ones were bred.

PSO. The PSO began with a population of 10 randomly created networks. The weights and biases were initialised in the same manner as the EA. The king value was a random value between 1 and 3. The self adaptive parameters acted as the velocity for each weight and bias, and was initialised to 0. At each time step the position of each network was updated by

$$x_{i,j}[t+1] = x_{i,j}[t] + v_{i,j}[t] \qquad (3)$$

Where:

t = time, v = velocity, x = the position of i in the dimension j.
The velocity was also updated at each time-step:

$$v_{i,j}[t+1] = v_{i,j}[t] + (c_1 \times \text{rand}() \times (pbestx_{i,j} - x_{i,j})) + (c_2 \times \text{rand}() \times (gbestx_{i,j} - x_{i,j})) \quad (4)$$

Where:

c_1 & c_2 = constants, $rand()$ = random number between 0 and 1,

$pbest$ = the network's best position, $gbest$ = the population's best position.

For our experimentation $c1$ and $c2$ were both set to 2 [13]. The velocities were limited to between -0.1 and 0.1, to prevent the values from large oscillations. Each network played every other network in the population to determine the first global best. Each iteration, each solution's position is updated, they then play every other network in the population to determine the best network of that iteration, which then plays the global best to see if a improvement had been found. Each network then plays their own local best to see if an improvement had been found.

4   Results

4.1   Training Methods

The Effect of the Amount of Plies Used in Training. To test the effect of the amount of minimax plies used in training on the resulting networks quality, we trained multiple networks, using 2 and 4 plies in their training games, for PSO and EA against an opponent with a simple piece differential as a utility function (subtracting one player's pieces from the other to determine game state). The training was set so that each network would have around 8 hours to train on our machine, which was 500 generations for EA 4 plies, 5000 generations for EA 2 plies, 300 iterations for PSO 4 plies, and 3000 iterations for PSO 2 plies. After this the 2 and 4 plies trained network of each method played against the other, as seen below (Tables 1, 2, 3, 4, 5 and 6).

Table 1. Results from 200 sets of games on 4 plies for the EA trained networks

	Result (for 2 plies)		
Playing as	Win	Loss	Draw
White	40.0%	32.0%	28.0%
Black	28.0%	47.0%	25.0%

Table 2. Results from 100 sets of games on 6 plies for the EA trained networks

	Result (for 2 plies)		
Playing as	Win	Loss	Draw
White	28.0%	42.0%	30.0%
Black	31.0%	32.0%	37.0%

Table 3. Results from 50 sets of games on 8 plies for the EA trained networks

	Result (for 2 plies)		
Playing as	Win	Loss	Draw
White	26.0%	18.0%	56.0%
Black	26.0%	34.0%	40.0%

Table 4. Results from 200 sets of games on 4 plies for the PSO trained networks

	Result (for 2 plies)		
Playing as	Win	Loss	Draw
White	52.5%	19.5%	28.0%
Black	60.0%	17.5%	22.5%

Table 5. Results from 100 sets of games on 6 plies for the PSO trained networks

	Result (for 2 plies)		
Playing as	Win	Loss	Draw
White	7.0%	72.0%	21.0%
Black	33.0%	3.0%	64.0%

Table 6. Results from 50 sets of games on 8 plies for the PSO trained networks

	Result (for 2 plies)		
Playing as	Win	Loss	Draw
White	24.0%	30.0%	46.0%
Black	68.0%	16.0%	16.0%

As we can see for both training methods, training on both 2 plies and 4 plies produced capable networks. The EA trained networks (Tables 1, 2 and 3) appear to be similar in ability, however the 4 plies trained network seems to slightly outperform the 2 plies trained network, as it wins more games. For the PSO trained networks (Tables 4, 5 and 6) the 2 plies trained network outperforms the 4 plies trained network, and it is far less close than EA.

Whilst both are capable of training accurate networks, it appears that 2 plies is slightly better, as it is able to complete 10 times the generations 4 plies can. This increase in generations allows more mutations or movements to occur, allowing a larger amount of the search space to be explored. Due to the vast amount of weights and biases within the network the search space is extremely large, and this greater amount of exploration helps to find better solutions.

An interesting observation is for certain amount of plies the result seems to be determined by the side the network is playing as. For example in Table 5 we see how the 2 plies network loses the majority of games when playing as white, however it wins the majority playing as black. Whilst this may seem erroneous, it only occurs on trained networks, and by manually checking we were able to observe the fact it is not due to them playing as black, but rather having the second move. If the game started on blacks turn it would lose the majority of the games. It is difficult to say why the network behaves this way, however we do not believe it is important, as we later show that it is able to play well as both black and white, against both EA trained networks and piece differential.

EA vs PSO. Knowing that the EA trained 4 plies and PSO trained 2 plies were the best performing networks, we then played them against each other, with each network searching to the same amount of plies, allowing us to compare the suitability of each training method for this task (Tables 7, 8 and 9).

Table 7. Results from 200 sets of games on 4 plies

	Result (for PSO)		
Playing as	Win	Loss	Draw
White	23.5%	0.5%	76.0%
Black	29.0%	31.5%	39.5%

Table 8. Results from 100 sets of games on 6 plies

	Result (for PSO)		
Playing as	Win	Loss	Draw
White	19.0%	5.0%	76.0%
Black	100.0%	0.0%	0.0%

Table 9. Results from 50 sets of games on 8 plies

	Result (for PSO)		
Playing as	Win	Loss	Draw
White	54.0%	8.0%	38.0%
Black	42.0%	6.0%	52.0%

As we can see, the PSO trained network wins more games than the EA trained network across a range of plies.

An explanation for this is the way in which each method explores the search space. The search space is particularly large for this problem, as each weight and bias represents another dimension, and there are 6484 weights and biases within the current network architecture. As the EA used single parent reproduction it has a higher risk of getting stuck in local minima. This is because during training a dominant network will appear, and over time the population will be filled the mutated ancestors of this network. This greatly limits the explored space. Whilst PSO is still at risk of getting stuck in local minima, as all solutions have to converge on the global best a greater amount of the search space will be explored, meaning if the global best is in a local minima during convergence it is likely a better solution will be found, however this still may not be the global minima.

It is important to note that whilst the fully trained 2 plies PSO trained network outperformed the 4 plies EA trained network it does not necessarily prove that PSO is superior. Both of these training methods are susceptible to becoming stuck in local minima [17], and as such the initial starting positions has a great impact on the end result of the network. To counteract this, 3 networks were trained for each parameter, using the best one within the testing, however we still cannot be certain PSO outperforms EA. In order to be able to increase our confidence in these results we could train more networks, as this would increase the range of starting positions. This was not possible due to time restrictions. It would also be beneficial to increase population size for both training methods, as this would increase the state space explored, however this would also increase the time taken training.

4.2 Neural Networks as Evaluation Functions

Neural Network Vs Piece Differential on Fixed Amount of Plies. In order to test the accuracy of the 2 plies PSO network, and to see if it was more accurate than a piece differential, it played a number of games against the piece differential, as both white and black, on varying amounts of plies.

In this experiment we see the average score for the first time. A win for white was awarded a score of 1, a win for black was awarded -1, and a draw was given a score dependant on the remaining pieces. This score was the total piece differential at the end of the game, with a king given a value of 2, which was divided by 12 and limited to between 1 and −1. The closer to 1 the better the draw was for white, and the closer to −1 the better for black. Using the results from each individual game we can calculate the average score (Tables 10, 11 and 12).

Table 10. Results from 200 sets of games on 4 plies

	Result (for the network)			
Playing as	Win	Loss	Draw	Avg score
White	63.5%	7.0%	29.5%	0.55
Black	53.0%	16.0%	31.0%	−0.363333

Table 11. Results from 100 sets of games on 6 plies

	Result (for the network)			
Playing as	Win	Loss	Draw	Avg score
White	72.0%	5.0%	23.0%	0.6775
Black	75.0%	7.0%	18.0%	−0.675

Table 12. Results from 50 sets of games on 8 plies

	Result (for the network)			
Playing as	Win	Loss	Draw	Avg score
White	72.0%	4.0%	24.0%	0.686667
Black	72.0%	2.0%	26.0%	−0.711667

As we can see the network is able to consistently win more games than the piece differential across a range of plies. This shows that the training was successful in producing networks that could more accurately score game states within minimax than a piece differential.

An interesting note is the higher the amount of plies, the more successful the network appears to be. A possible explanation for this is the networks ability to process the spacial characteristics of the board, enabling it to recognise an advantageous position where the piece differential will not. By recognising this position earlier, due to the higher amount of plies, the system is able to attempt to move the game towards this position earlier, thus increasing the chances of the advantageous position being reached, and from it a better result being achieved.

Neural Network vs Piece Differential Using Iterative Deepening. As the network took longer to evaluate states than the piece differential, we played them against each other, using iterative deepening instead of a fixed ply. By testing this we can ensure the network is superior, and see whether the longer time taken to evaluate states is counteracted by the higher accuracy of the network.

The games were only played 8 times, as the CPU used has 8 cores, and as such playing more games simultaneously would decrease the depth reached within the time limit. Each player would have to share the core, decreasing the CPU time it would have, and thus the depth it would reach (Tables 13, 14 and 15).

Table 13. Results from 1 s of iterative deepening

	Result (for the network)		
Playing as	Win	Loss	Draw
White	4	1	3
Black	5	0	3

Table 14. Results from 3 s of iterative deepening

	Result (for the network)		
Playing as	Win	Loss	Draw
White	2	1	5
Black	5	0	3

Table 15. Results from 5 s of iterative deepening

	Result (for the network)		
Playing as	Win	Loss	Draw
White	4	1	3
Black	3	0	5

Whilst piece differential was able to consistently reach 2–4 higher plies, it was not able to outperform the network, with the network winning the majority of the games. We can see that the results were closer than the games played whilst searching to a fixed depth, due to the increased amount of plies reached by the piece differential. This shows that not only are neural networks suitable as evaluation functions, they can significantly outperform simpler evaluation functions, such as a piece differential. It is worth noting that this is a smaller sample size, and as such is less accurate than some of the higher game tests, however it is enough to show that the trained network is superior to the piece differential.

4.3 Game Theory

The Effect of the Amount of Plies Searched to. In order to test the effect of the amount of plies searched to on the performance of the system, two piece differentials played against each other, set to search to a different amount of plies.

As we can see from the results below, increasing the number of plies searched to at any depth significantly increased the performance of the system, allowing it to win the majority of the games (Tables 16, 17, 18, 19 and 20).

Table 16. Playing 200 games of 4 plies against 2 plies

	Result for 4 plies			
Playing as	Win	Loss	Draw	Avg score
White	80.0 %	2.5%	17.5%	0.77625
Black	77.5%	3.0%	19.5%	−0.753333

Table 17. Playing 100 games of 6 plies against 2 plies

	Result for 6 plies			
Playing as	Win	Loss	Draw	Avg score
White	86.0%	1.0%	13.0%	0.848333
Black	89.0%	0.0%	11.0%	−0.896667

Table 18. Playing 100 games of 6 plies against 4 plies

	Result for 6 plies			
Playing as	Win	Loss	Draw	Avg score
White	56.0%	3.0%	41.0%	0.533333
Black	54.0%	6.0%	40.0%	−0.490833

Table 19. Playing 50 games of 8 plies against 4 plies

	Result for 8 plies			
Playing as	Win	Loss	Draw	Avg score
White	70.0%	4.0%	26.0%	0.66
Black	72.0%	2.0%	26.0%	−0.676667

Table 20. Playing 50 games of 8 plies against 6 plies

	Result for 8 plies			
Playing as	Win	Loss	Draw	Avg score
White	40.0%	12.0%	48.0%	0.293333
Black	60.0%	6.0%	34.0%	−0.523333

We can also see that the measure of success is not the absolute difference between the number of plies, but the relative difference between them. For example in Table 16 we see 4 plies playing against 2 plies. The results are far closer to 8 plies against 4 plies (Table 19), than 8 plies against 6 plies (Table 20). Another example is playing 6 plies against 2 plies (Table 17) scored significantly higher than 8 plies against 4 plies, despite them having the same difference. An explanation for this is the diminishing returns of higher amount of plies searches. Increasing the plies allows the search to be more accurate, and closer to the optimal move. Whilst a higher number of plies will always increase the accuracy of the search, there becomes a point where this becomes negligible. For example consider a minimax search that is capable of searching until terminal nodes, removing the need for an evaluation function. If we had another search that was capable of reaching 2 plies before the terminal nodes it is likely that these searches would almost always return the same move to take, especially when combined with an accurate evaluation function. This would cause the systems to be extremely close in performance. The increase in accuracy searching 4 plies instead of 2 plies offers is far more significant than 8 plies instead of 6 plies, and it is this accuracy of the search which determines the performance.

4.4   Measuring the Performance of the System

The site used for evaluation by Chellapilla and Fogel is largely inactive, and it seems to be broken or hacked, as players have ratings of well above the expected value for human players. The current highest rating at the time of writing is 4603, and according to Chellapilla and Fogel's paper [6] 2400+ is the equivalent of a senior master, the highest ranking a checkers player can achieve. At the time of its retirement Chinook had a rating of 2814. It is worth noting here that the issue of the ELO system is that it relies on the skill level of the population, as it is the player's skill vs the skill of others in the population. This means that a professional player's rating would not be same on a website, as the skill level would not be the same on the website as it would in professional play. This means that we cannot accurately translate ratings from one site to another, and as such we cannot compare this system directly to Chellapilla and Fogel's.

Each of the following tests used the previously found best settings for the system, that is using the 2 plies PSO trained network, 120 s of iterative deepening (the maximum time for a move in most common rule variants), alpha-beta pruning enabled, and an endgame database, unless otherwise specified.

Human Players. Whilst we were unable to directly play the system against expert players we were able to have human players play the system in order to show its performance. The person was asked to estimate how many games of checkers they had played, and the rules were explained to them. Once the player was informed of all the rules, they played a total of 4 games, 2 as white and 2 as black. In the results players are grouped by their experience with checkers, and each game of the four they played is recorded as a separate result in the Table 21.

Table 21. System playing against non-expert players

	Result (for system)		
Previous amount of games played	Win	Loss	Draw
0–9	20	0	0
10–99	16	0	0
100+	8	0	0

The system had 3 s of iterative deepening search, alpha-beta pruning enabled and no EGDB access. The plan was to increase the search time and enable the EGDB if the player beat this version of the system. The reasoning behind this was not wanting to waste the participants' time, as a game on 2 min per move will easily take over an hour. However, none of the players were able to beat this "easy" version of the system. This demonstrates the high performance of the system, and shows that it surpassed the minimum aim of being competitive with non-expert human players.

Checker Cruncher. The website Checker Cruncher [1] provides a number of problems, which are from recorded games where the player who is moving has the opportunity to win the game based upon the next move. It has over 14,000 problems, each with their own rating. These ratings are calculated through the Glicko system, which is an improved ELO system. Whilst the scores are not directly comparable, it still allows us to provide context to the performance of the system. Alongside this the website has a list of members, each with a rating attached, based upon the challenges they had attempted. This allows us to compare the system to these users, however we cannot be certain of the actual skill of these players, or if they also, like us, are using a game playing system.

In order to measure the system it was used to select the move for multiple problems, based upon their rating, starting at lower rated problems, and moving to harder ones. The system was set to use alpha-beta pruning, 5 s of iterative deepening, and the EGDB. Ratings are correct as of time of writing. In order, the system attempted problems (Glicko rating is to the right of them):

1. 1353 (650.4)
2. 5151 (1031.9)
3. 6922 (1218.1)
4. 505 (1471.5)
5. 12542 (1653.2)
6. 5835 (1740.0)
7. 11136 (1788.5)
8. 1454 (1833.5)
9. 5797 (1927.5)
10. 7488 (1996.4)
11. 8465 (2040.4)
12. 6249 (2075.7)
13. 14112 (2125.8)

The system was able to pick the correct move in all problems except problems 11136 and 14112. Within these it chose good moves, but the website said there were better ones. Increasing the iterative deepening time to 2 min allowed the system to select the correct move for both problems. It is a possibility that the system will not perform as well when recreating these results, due to the random selecting of moves with equal scores, however there is only one problem we know this occurs on, which is 14112, where it can select either the correct move or the good move mentioned previously.

14112 is currently the hardest rated problem on the site, and this goes to show the power of the system. Whilst the accuracy of these ratings cannot be guaranteed, it is still a good indicator of the performance of the system, and that it is capable of playing checkers to a high level. It is worth noting only 3 of these problems (5151, 505 and 7488) used the EGDB, so the other results come from the network alone.

The problems were chosen before creating an account, and were hard coded on the system. To get a rating on the site you have to attempt random problems, meaning it was not possible to get a rating on the site. However by working out the Glicko scores manually we get a new Glicko rating for the system of 1984. This however is a conservative estimate, as it has won every game, and as such the true rating may be much higher, and would rise further with repeated solves of harder problems. This rating would place the system at rank 1 on the Checker Cruncher leader board, out of 374 players. At the time of writing the current leader has a Glicko rating of 1972.1.

Chinook. Chinook is available to play online [18]. The online version of Chinook appears to be slightly downgraded, only using a 6 piece EGDB, however being able to draw with it would be a considerable achievement. It offers 3 difficulty settings (novice, amateur or intermediate), each one increasing the amount of CPU time Chinook uses to chose its move. In order to test the performance of the system it was played it against Chinook, using the full EGDB, 2 min of iterative deepening, and alpha-beta pruning enabled. In order to play them against each other a game was started on our system with the network playing as white, and another game on Chinook with us playing as black (white on our board representation). We would then relay the move made by the system into Chinook, and enter its reply into our system. Starting at novice and moving through the difficulties the system was able to draw at every difficulty, whilst playing as black (white on our system), and starting from the default position. One thing to note is the online version does not seem to draw properly, with games looping forever, however games were ended when a draw was registered on our system.

Whilst we cannot know how inferior this online version is compared to the full system (if at all), this is still a significant achievement, as it goes to show the power of the system, as well as the suitability of the training methods, being able to draw with such a system, with a far smaller development time.

5   Conclusion

Within this paper we have found that Particle Swarm Optimisation and Evolutionary Algorithms are both capable training methods for neural networks within a game playing system. We have found that 2 and 4 plies are both suitable amount of plies to train on. We have found that Particle Swarm Optimisation appears to be more suited to training ANNs with a large amount of weights and biases, due to it being less susceptible to local minima than EA. Additionally we have shown that this fully trained network can outperform a piece differential, both on a fixed amount of plies and when using iterative deepening, showing that although it is slower, the decrease in speed is counteracted by its higher accuracy as an evaluation function.

We have shown that increasing the number of plies will significantly increase the performance of the system, especially towards the lower number of plies. However the measure of success is the relative difference in the amount of plies, not the absolute difference.

Finally we have externally validated this system, against human players, on a checkers problem website "Checker Cruncher", and the well-known Chinook AI. The system showed that it was extremely capable against humans, winning every single game, was able to perform better than all humans on the checker problem site, and was even able to force Chinook into a draw, the best result obtainable against that system.

This type of learning has many real world applications, for example within the domains of science, economics and the environment. Whilst within this project we have applied it to playing a game, it can be translated to other problems, where the rules are known, but we are unable to produce an optimal solution. Alongside producing solutions that can be superior to solutions produced by humans, this type of learning can even be applied to problems that aren't yet fully understood.

We have shown that Evolutionary Algorithms and Particle Swarm Optimisation can be successfully applied to Artificial Neural Networks to enable high-quality game-playing systems.

References

1. Checker Cruncher. https://www.checkercruncher.com/
2. Go Ratings. https://www.goratings.org/en/
3. ItsYourTurn.com - Help Page. https://www.itsyourturn.com/t_helptopic2030.html
4. Boonzaaier, D.: Training neural networks for checkers. Ph.D. thesis (2017)
5. Campbell, M., Hoane, A.J., Hsu, F.H.: Deep Blue. Artif. Intell. **134**(1–2), 57–83 (2002). https://doi.org/10.1016/S0004-3702(01)00129-1
6. Chellapilla, K., Fogel, D.B.: Evolving neural networks to play checkers without relying on expert knowledge. IEEE Trans. Neural Netw. **10**(6), 1382–1391 (1999). https://doi.org/10.1109/72.809083
7. Franken, N., Engelbrecht, A.P.: Comparing PSO structures to learn the game of checkers from zero knowledge. In: 2003 Congress on Evolutionary Computation, CEC 2003 - Proceedings (2003). https://doi.org/10.1109/CEC.2003.1299580
8. Hsu, F.H., Campbell, M., Hoane, J.: Deep Blue system overview. In: Proceedings of the 9th International Conference on Supercomputing, ICS 1995, pp. 240–244 (1995). https://doi.org/10.1145/224538.224567
9. Kennedy, J., Eberhart, R.: Particle swarm optimization. In: Proceedings of ICNN 1995 - International Conference on Neural Networks, vol. 4, pp. 1942–1948 (November 1995). https://doi.org/10.1109/ICNN.1995.488968
10. Knuth, D.E., Moore, R.W.: An analysis of alpha-beta pruning. Artif. Intell. **6**(4), 293–326 (1975). https://doi.org/10.1016/0004-3702(75)90019-3
11. Korf, R.E.: Depth-first iterative-deepening. An optimal admissible tree search. Artif. Intell. **27**(1), 97–109 (1985). https://doi.org/10.1016/0004-3702(85)90084-0
12. Messerschmidt, L.: Using particle swarm optimization to evolve two-player game agents (2005)
13. Poli, R., Kennedy, J., Blackwell, T.: Particle swarm optimization. Swarm Intell. **1**, 33–57 (2007). https://doi.org/10.1007/s11721-007-0002-0
14. Schaeffer, J., et al.: Checkers is solved. Science **317**(5844), 1518–1522 (2007). https://doi.org/10.1126/science.1144079
15. Silver, D., et al.: Mastering the game of Go without human knowledge. Nature (2017). https://doi.org/10.1038/nature24270
16. Sousa-Ferreira, I., Sousa, D.: A review of velocity-type PSO variants. J. Algorithms Comput. Technol. (2017). https://doi.org/10.1177/1748301816665021
17. Taherdangkoo, M., Paziresh, M., Yazdi, M., Bagheri, M.H.: An efficient algorithm for function optimization: modified stem cells algorithm. Cent. Eur. J. Eng., pp. 36–50. (2013). https://doi.org/10.2478/s13531-012-0047-8
18. University of Alberta: Play Chinook. https://webdocs.cs.ualberta.ca/~chinook/play/

A Novel Outlook on Feature Selection as a Multi-objective Problem

Pietro Barbiero[1] , Evelyne Lutton[2] , Giovanni Squillero[1] ,
and AlbertoTonda[2]([envelope])

[1] Politecnico di Torino, Torino, Italy
pietro.barbiero@studenti.polito.it, giovanni.squillero@polito.it
[2] UMR 782, Université Paris-Saclay, INRA, AgroParisTech,
Thiverval-Grignon, France
{evelyne.lutton,alberto.tonda}@inra.fr

Abstract. *Feature selection* is the process of choosing, or removing, features to obtain the most informative feature subset of minimal size. Such subsets are used to improve performance of machine learning algorithms and enable human understanding of the results. Approaches to feature selection in literature exploit several optimization algorithms. Multi-objective methods also have been proposed, minimizing at the same time the number of features and the error. While most approaches assess error resorting to the average of a stochastic K-fold cross-validation, comparing averages might be misleading. In this paper, we show how feature subsets with different average error might in fact be non-separable when compared using a statistical test. Following this idea, clusters of non-separable optimal feature subsets are identified. The performance in feature selection can thus be evaluated by verifying how many of these optimal feature subsets an algorithm is able to identify. We thus propose a multi-objective optimization approach to feature selection, EvoFS, with the objectives to i. minimize feature subset size, ii. minimize test error on a 10-fold cross-validation using a specific classifier, iii. maximize the analysis of variance value of the lowest-performing feature in the set. Experiments on classification datasets whose feature subsets can be exhaustively evaluated show that our approach is able to always find the best feature subsets. Further experiments on a high-dimensional classification dataset, that cannot be exhaustively analyzed, show that our approach is able to find more optimal feature subsets than state-of-the-art feature selection algorithms.

Keywords: Feature selection · Machine learning · Multi-objective optimization · Evolutionary algorithms · Multi-objective evolutionary algorithms

1 Introduction

The field of *machine learning* (ML) deals with algorithms producing predictive models, that are able to improve their performance over time, given an increasing amount of data. *Supervised* ML, a category including the notable examples

© Springer Nature Switzerland AG 2020
L. Idoumghar et al. (Eds.): EA 2019, LNCS 12052, pp. 68–81, 2020.
https://doi.org/10.1007/978-3-030-45715-0_6

of classification and regression, defines a set of problems for which training data is labeled. In ML terminology, data is organized in *samples*, each reporting measurements over a set of *features*; in other terms, samples can be seen as the rows in a dataset, while features can be seen as the columns. The aim of a supervised ML algorithm is to find relationships between features that can reliably predict the value of the *target*, a specific feature in the problem.

While ML algorithms can often be successful at a given task, they might face issues when dealing with large number of features, as an increase in dimensionality creates a corresponding increase in the search space of combinations of features to be explored. As the search space grows, it becomes harder for the ML algorithms to find good optima. Not only, but even when the results are satisfying, very often the predictive models obtained are *black boxes*, that cannot be interpreted by humans. Selecting the features involved in a problem can help not only to reduce the search space for the ML algorithms to explore, but also to make the models more human-readable.

Specialized literature on feature selection shows different approaches to scoring feature subsets, ranging from mutual information to analysis of variance. Evolutionary algorithms (EAs) have been successfully used for feature selection [1,2], even with multi-objective approaches, attempting to both minimize the number of features in a subset, and minimize error [3,4], with recent applications ranging from face recognition [5] to medicine [6].

Most of the proposed feature selection algorithms relying upon classifiers or regressors to obtain an evaluation of a feature subset, exploit a K-fold cross-validation to better assess average error. As this is a stochastic process that will return K error values, a K-fold cross-validation can be assimilated to sampling an unknown probability distribution K times; thus, comparing feature subsets on just their average error might be misleading. A more formally correct approach would be to assess the likelihood that the two sets of K error values have been sampled from two different distributions, using a statistical test. This statistical comparison can uncover to the existence of clusters of feature subsets whose performance is non-separable, and can thus be considered equally optimal. To the best of the author's knowledge, this analysis is usually not considered in feature selection literature.

In this work, we propose a novel approach to multi-objective feature selection, that we call EvoFS. The objectives to be optimized are: i. minimize feature subset size, ii. minimize test error on a 10-fold cross-validation using a specific classifier, iii. maximize the analysis of variance value of the lowest-performing feature in the set. The third objective improves human understanding of the results, as it pushes for feature subsets where all relationships between single features and the target are significant.

Taking into account the statistical considerations of non-separability of performance for feature subsets, it is possible to better assess experimental results. Experimental evaluations on simple datasets, that can be completely analyzed, show that the proposed approach reliably uncovers more feature subsets inside the clusters of the non-separable, optimal ones, when compared to state-of-the-art feature selection algorithms. Further experiments on a high-dimensional dataset confirm the previous results.

2   Background

2.1   Machine Learning

Given a set of samples $\bar{x}$, and a set of corresponding values for a target $\bar{y}$, generated by an unknown function f, a supervised ML algorithm has the objective to learn an approximation $\hat{f}$, so that the values predicted by $\hat{f}$ for $\bar{x}$ have the least possible error with respect to $\bar{y}$.

2.2   Feature Selection

In ML, feature selection is the process of choosing (or eliminating) features from a dataset, reducing them to the minimal, most informative subset. Removing information might, at first glance, seem detrimental for the performance of ML algorithms: however, certain features might just add noise; or they might be redundant, for example being heavily correlated with others; and finally, eliminating features reduces the search space that ML algorithms have to explore, facilitating the task of finding effective models.

Besides improving the performance of ML algorithms (not only in terms of computation time but also regarding precision of results [7]), feature selection can also be used to reduce information and ultimately make it human-readable. For example, while reviewing the contributions of 1,000 different variables in a problem is impossible for human experts, a selection of 10 highly-informative features can usually be analyzed, even if relevant parts of the information are removed. This is particularly useful when dealing with genomic or other high-dimensional data [8]. More generally, feature selection is one facet of dimensionality reduction, which is an important domain in the field of data visualization [9].

Feature selection can be performed using various approaches [10], simple ones consist in filtering the features according to a criterion (often based on statistical tests), or in using recursive procedures (forward or backwards) to eliminate redundant features [11,12]. Subset selection methods are more complex and rely on the definition of a quality measurement of the subset. The problem is thus turned into an optimization one: selecting the best subset of features that maximizes an objective function (usually a "goodness-of-fit" combined with a regularization term, including a penalty for a large number of variables [10,13]). Several single-objective EAs have been proposed, exploiting similar scores for the fitness function [1,2]. Finally, feature construction and space dimensionality is another way to reduce information. Subsets made of combinations of features are built for a better representation of the dataset (dimensionality reduction methods, principal component analysis for instance).

Given a candidate subset of features, evaluating its efficacy is not trivial. Ideally, what would need to be measured is the *content of information* of the feature subset, and several metrics have been proposed to assess it in literature: for example *mutual information* [14] or *analysis of variance* [15]. In practice, however, even the most popular metrics can only assess part of the information content of a feature subset, as taking into account the contribution of non-linear combinations of features is too computationally expensive.

A different way to assess efficacy for a feature subset is using it as input of a ML algorithm, and evaluate the difference in performance compared with the same algorithm, using all features, or a different feature subset. To avoid issues with overfitting, a K-fold cross-validation can be used, obtaining an average of its performance (for example, classification accuracy) on the test folds. As the cross-validation procedure is stochastic, comparing two feature subsets on just their average performance on test folds is not enough, because the variance of the results is not taken into account. A more robust approach is to consider the K performance results on test folds of the two feature subsets as samples drawn from two probability distributions, and exploit a statistical test to assess the likelihood that the two sets of samples are drawn from different distributions. If the two sets of samples are separable below an arbitrary confidence threshold, for example $p < 0.05$, the feature subset with the best average performance can be considered better than the other. The main issue of this methodology is that it is sometimes impossible, with the available data, to separate the performance of different feature subsets.

2.3 Multi-objective Evolutionary Algorithms and Feature Selection

Multi-objective optimization algorithms aim at finding the best compromises between conflicting criteria, ultimately delivering a set of non-comparable, non-dominated solutions to the users. Evolutionary Algorithms (EAs) currently represent the state-of-the-art in the field, with the Multi-Objective EA (MOEA), Non-Sorting Genetic Algorithm II (NSGA-II) [16] being one of the most widely adopted for real-world applications.

Given their effectiveness, it is not surprising that MOEAs have been already applied to feature selection problems, where the conflicting objectives are usually: i. minimizing the number of features and ii. maximizing a quality metric for a feature subset. In [3] the authors apply NSGA-II for feature selection. In [4], differential evolution is used instead. MOEA approaches to feature selection have been recently applied to facial recognition [5] and medical imaging [6].

3   Proposed Approach

We propose a novel approach to feature selection in ML, framing it as a multi-objective problem with three aims: i. minimizing the number of features; ii. minimizing error on a cross-validation; iii. maximizing mutual information content between each feature and the target. Feature selection can be seen as finding the best compromises between the number of features considered and the final result for a ML algorithm. However, assessing the effectiveness of the selected features for the problem is far from trivial, and only indirect metrics are available.

It must be noted that analyzing all feature subsets for a given dataset is often impossible, as the total number of feature subsets of dimension d for a dataset with F features is:

$$\sum_{d=1}^{F} \binom{F}{d} = 2^F \tag{1}$$

3.1   Individual Representation

Individuals represent feature subsets, and are internally stored as simple bit-strings of size equal to the number of features in the original dataset. A '1' in the i-th position of an individual means that the corresponding i-th feature is included in the subset; a '0' indicates that the i-th feature is not included in the subset.

3.2   Fitness Functions

The first objective in the proposed approach is to minimize the number of features included in a subset:

$$O_1 = \sum_{i=1}^{F} I(i) \tag{2}$$

where I is an individual represented as a bit-string, $I(i)$ indicates the bit in i-th position, and F is the number of features in the problem, also corresponding to the size of an individual.

The second objective assesses the effectiveness of a candidate feature subset for a specific problem, through a K-fold cross-validation, a procedure where training data is divided into K parts, termed *folds*, that are alternatively used for training and test. This objective can be stated as:

$$O_2 = \frac{1}{K} \sum_{i=1}^{K} L_{k(i)} \tag{3}$$

where K is the number of folds; $k(i)$ is the i-th fold. $L_{k(i)}$ is defined as:

$$L_{k(i)} = L(y_{k(i)}, \hat{g}^{-k(i)}(x_{k(i)})) \tag{4}$$

where L is an error function, evaluating the differences between the values predicted by $\hat{g}^{-k(i)}$ and the known values $y_{k(i)}$; $\hat{g}^{-k(i)}$ is the function learned by a ML algorithm, trained on all data, except fold $k(i)$; in general, $\hat{g}$ is always considered to be an approximation of the real function g that generated the known values of y; $y_{k(i)}$ and $x_{k(i)}$ are the known values of the target and the corresponding features for samples in $k(i)$, respectively. The error measured by L is averaged over the K folds to obtain the final value of O_2.

Finally, the third objective is a proxy for human readability of the candidate feature subset. Using the one-way Analysis of Variance (ANOVA) F-value procedure [17], that captures univariate relationships between a feature and the target. Indeed, the F-value of the i-th feature ϕ_i can be interpreted as the proportion of variance explained by the feature to the total variance in the data. If we make a reasonable assumption that an higher amount of explained variance may correspond to a higher discriminating capability, then we can rank features according to their ϕ, where the best feature will have the highest value. Finally, for each subset of features (i.e. the candidate solution), the third fitness objective is function of the worst ϕ in the feature subset:

$$O_3 = \frac{1}{\min(\phi_0, \phi_1, ..., \phi_f)} \tag{5}$$

where f is the number of features in the subset. This objective will force the evolutionary process to drop feature sets containing at least one variable whose univariate contribution is negligible. In fact, ML classifiers as well as other automatic FS algorithms (such as RFE) risk to retain a dramatic amount of features which are not a true causative source of the observed phenomenon (a.k.a. false positives). However, they are often selected as they might be slightly correlated with the target, providing a minor contribution to the classification accuracy.

Taking into account Eqs. 2, 3 and 5, the multi-objective optimization problem can be described as:

$$argmin(O_1, O_2, O_3) \tag{6}$$

4   Experimental Results

The experiments presented in this work deal with classification only, due to the greater availability of high-dimensional classification datasets in the public domain; but the proposed methodology can also be straightforwardly applied to regression problems. The experimental evaluation of the proposed approach is divided into two parts. Firstly, datasets that have a relatively low dimensionality (9–18 features) are analyzed: as all feature subsets for these datasets can be explored exhaustively, we can assess whether there is actually a single best feature subset, and whether different methodologies are able to find it. In a second batch of experiments, the proposed methodology tackles an artificial dataset with high dimensionality (500 features), that cannot be analyzed exhaustively, but whose characteristics are completely known.

4.1   Experimental Setup

For the following experiments, the error function L (see Eq. 4) is classification error, an established quality metric for classifiers, simply defined as the ratio between incorrect predictions and total predictions. The closer classification error is to 0, the higher the quality of the predictive model. The classifier used to learn $\hat{f}$ in O_2 is Logistic Regression [18], a popular algorithm of proved effectiveness.

The MOEA selected for the experiments is NSGA-II [16], that currently represents the state of the art for multi-objective optimization with up to three objectives. After preliminary evaluations, NSGA-II's parameters are set to: $\mu = 100$, $\lambda = 100$, probability of crossover $p_c = 0.9$, probability of mutation $p_m = 1/l$, where l is the length of an individual, and a stop condition based on the maximum number of generations, set to 200.

The proposed approach, termed *EvoFS* in tables and figures, is compared against three popular state-of-the-art feature selection methods: recursive feature elimination (RFE) [19], that uses a classifier to score a feature set, then iteratively removes the lowest-performing feature and scores the subset again; greedy forward selection, that greedily adds features to a subset, using either their mutual information (MI) [14], or analysis of variance (ANOVA) [15] scores. All these methods need the user to specify the number of features to be selected, so in the experiments they have been called once for every possible size of feature subset in the problem, to have a fair comparison.

As previously stated, comparing the effectiveness of two feature subsets for classification using the error function L is not trivial, due to possible random effects in the classifier's training process, or in the way the training/test split of the data is performed. Randomly dividing the data in K folds and performing a K-fold cross-validation can help obtain a better average for L, but introduces further stochasticity in the process. When comparing results in this work, we consider the outcome of a K-fold cross-validation as K separate samples coming from an unknown statistical distribution. We then compare the results of two feature subsets as if assessing the likelihood that their accuracy scores have equal means. As we cannot assume that the two distributions have the same standard deviation, we use a Welch's T-test [20] with an arbitrary but commonly accepted threshold for the p-value ($p < 0.05$). Such a statistical test assumes that samples are drawn from populations that are normal in shape. As pointed out in [21], this assumption is quite easy to meet for a wide range of practical distributions at a significance level $\alpha = 0.05$ and a sample size of $K \geq 5$. In the following, $K = 10$. We will use this procedure to isolate clusters of feature subsets that are non-separable for their classification error, and can thus be considered all equally optimal with regards to this metric. For each considered dataset, running times for all algorithms are reported in Table 2.

All the code in the experiments has been implemented in Python v3, using the modules `scikit-learn` [22] for all ML and feature selection algorithms, `openml` [23] for accessing the datasets in the OpenML repository, and `inspyred` [24] for NSGA-II. The scripts are freely available in a Bitbucket repository[1]. Experiments have been run on a consumer-end laptop[2].

---

[1] https://bitbucket.org/evomlteam/moea-feature-selection.
[2] Intel® Core™ i7-8750H 2.20 GHz, 8 GB RAM.

4.2 Simple Datasets

In a first set of experiments, simple datasets with a limited number of features are examined. The advantage of dealing with such datasets is that all their feature subsets can be enumerated and analyzed, a task that becomes impossible if dealing with hundreds or thousands of features. The datasets are freely accessible on the OpenML repository [25], and their characteristics are summarized in Table 1.

Table 1. Characteristics of the datasets used in the experiments.

Dataset name	Type	Features	Samples	Classes	Feature subsets
Diabetes [26]	Medical	9	768	2	512
Australian [27]	Credit scores	14	690	2	16,384
Vehicle [28]	Vehicle recognition	18	846	4	262,144
Madelon [29]	Artificial	500	4,400	2	10^{150}

In Figs. 1, 2 and 3, we show how many non-separable feature subset that have size lower or equal to the best performing one each algorithm was able to find: ideally, these are the ones that human users should be interested in. Then, for each algorithm, the position of each non-separable solution found is mapped into the exhaustive exploration of all feature subsets.

Figure 1 reports the results for the *diabetes* dataset. While all approaches are able to find feature subsets that are non-separable from the best ones, EvoFS finds the largest number. The same holds for the *australian* dataset, in Fig. 2, where notably RFE seems unable to find good solutions of small size. For *vehicle*, that features the largest search space so far, results reported in Fig. 3 show that, this time, RFE performs much better than the other two comparing methods, equalling the performance of EvoFS. Nevertheless, EvoFS is able to find a few non-separable solutions that are of smaller size than those uncovered by RFE.

An interesting general behavior that emerges from the plots, is that EvoFS is able to find non-separable feature subsets of lower size than the other algorithms. Notably, non-separable solutions of size larger than the best performing one are not included in its Pareto fronts.

4.3 High-Dimensional Datasets

The second set of experiments deals with a high-dimensional dataset, for which an exhaustive analysis of all feature sets is impossible. This datasets is artificial, taken from a classification competition focused on feature selection [29]. The datasets' characteristics are summarized in Table 1.

The target dataset, named *Madelon*, is an artificial dataset that can be procedurally generated, with a few informative features, several features that are linear combinations of the informative features, and a large number of deceiving

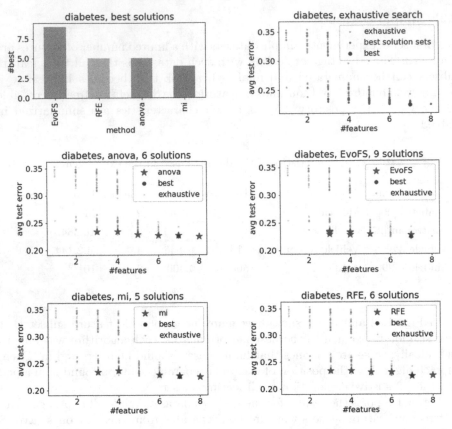

Fig. 1. (top left) Number of non-separable optimal feature subsets, of size less or equal to the one with the lowest error, found by each algorithm. **(top right)** All possible feature subsets for the dataset, identified exhaustively. In red, for each size, the ones that are non-separable. In green, the single feature subset with the lowest average error. **(middle-left to bottom-right)** Features subsets uncovered by the different approaches. (Color figure online)

features called *probes* [30]. For this work, we generated an instance of Madelon with the same parameters as the one featured in the competition [29]: 5 informative features, 15 linear combination features, 480 deceiving features/probes.

Figure 4 illustrates a summary of the results on the Madelon dataset. Remarkably, EvoFS is able to find a higher number of non-separable feature subsets having size lower or equal to the overall best solution. Moreover, EvoFS is also the only algorithm able to identify a non-separable solution of size 3, that includes only informative features (in positions 2, 3, 18).

While the greedy algorithms continue to be extremely fast on the high-dimensional dataset, it is noticeable from the running times reported in Table 2, how now RFE, with its iterative process, scales much worse than EvoFS.

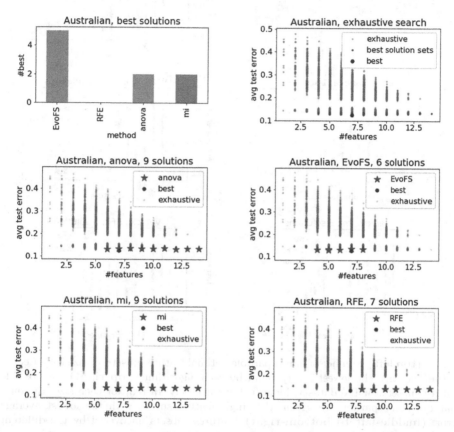

Fig. 2. (top left) Number of non-separable optimal feature subsets, of size less or equal to the one with the lowest error, found by each algorithm. **(top right)** All possible feature subsets for the dataset, identified exhaustively. In red, for each size, the ones that are non-separable. In green, the single feature subset with the lowest average error. **(middle-left to bottom-right)** Features subsets uncovered by the different approaches. (Color figure online)

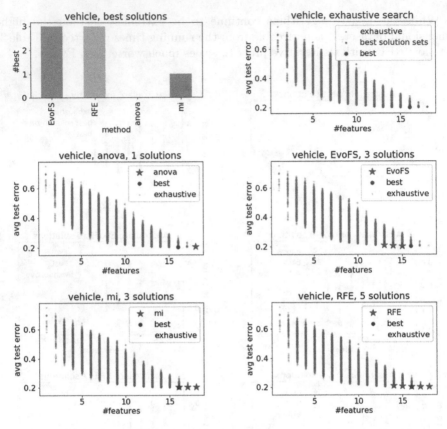

Fig. 3. (top left) Number of non-separable optimal feature subsets, of size less or equal to the one with the lowest error, found by each algorithm. **(top right)** All possible feature subsets for the dataset, identified exhaustively. In red, for each size, the ones that are non-separable. In green, the single feature subset with the lowest average error. **(middle-left to bottom-right)** Features subsets uncovered by the different approaches. (Color figure online)

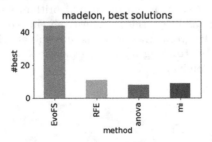

Fig. 4. Number of non-separable optimal feature subsets, of size less or equal to the one with the lowest error, found by each algorithm, on the *madelon* dataset.

Table 2. Running time (seconds) of the feature selection algorithms.

Dataset	EvoFS	Anova	MI	RFE
Diabetes	421.28 s	0.02 s	0.05 s	0.10 s
Australian	579.38 s	0.03 s	1.47 s	0.24 s
Vehicle	819.43 s	0.03 s	2.17 s	1.88 s
Madelon	3,549.57 s	0.05 s	12.97 s	18,925.29 s

5 Conclusions

Feature selection is an important task in ML, to obtain feature subsets of limited size that provide excellent performance. However, measuring performance is not trivial: commonly used metrics, such as the average error on a K-fold cross-validation, have been shown to mislead when comparing feature subsets that are, in fact, statistically non-separable. Using statistical tests, we uncover clusters of non-separable feature subsets in simple datasets, that can be exhaustively analyzed. Armed with this knowledge, we can then re-evaluate the performance of a feature selection methodology by estimating the number of optimal, non-separable feature subsets that the algorithm is able to discover.

The multi-objective feature selection algorithm we propose is shown able to find large numbers of feature subsets in such optimal clusters, when compared to other state-of-the-art algorithms in literature.

Future works will focus on further statistical comparisons with other evolutionary approaches to feature selection, and eventually introducing a human-interactive factor in the algorithm, in order to further promote human understanding of the results.

References

1. Cilia, N.D., De Stefano, C., Fontanella, F., Scotto di Freca, A.: Variable-length representation for EC-based feature selection in high-dimensional data. In: Kaufmann, P., Castillo, P.A. (eds.) EvoApplications 2019. LNCS, vol. 11454, pp. 325–340. Springer, Cham (2019). https://doi.org/10.1007/978-3-030-16692-2_22
2. Xue, B., Zhang, M., Browne, W.N., Yao, X.: A survey on evolutionary computation approaches to feature selection. IEEE Trans. Evol. Comput. **20**(4), 606–626 (2015)
3. Hamdani, T.M., Won, J.-M., Alimi, A.M., Karray, F.: Multi-objective feature selection with NSGA II. In: Beliczynski, B., Dzielinski, A., Iwanowski, M., Ribeiro, B. (eds.) ICANNGA 2007. LNCS, vol. 4431, pp. 240–247. Springer, Heidelberg (2007). https://doi.org/10.1007/978-3-540-71618-1_27
4. Xue, B., Fu, W., Zhang, M.: Multi-objective feature selection in classification: a differential evolution approach. In: Dick, G.G., et al. (eds.) SEAL 2014. LNCS, vol. 8886, pp. 516–528. Springer, Cham (2014). https://doi.org/10.1007/978-3-319-13563-2_44
5. Vignolo, L.D., Milone, D.H., Scharcanski, J.: Feature selection for face recognition based on multi-objective evolutionary wrappers. Expert Syst. Appl. **40**(13), 5077–5084 (2013)

6. Zhou, Z., Li, S., Qin, G., Folkert, M., Jiang, S., Wang, J.: Multi-objective based radiomic feature selection for lesion malignancy classification. IEEE J. Biomed. Health Inform. **24**, 194–204 (2019)

7. Fan, Y.J., Kamath, C.: On the selection of dimension reduction techniques for scientific applications (2012). 10.2172/1036865. part of the Annals of Information Systems book series (AOIS, volume 17)

8. Bermingham, M., et al.: Application of high-dimensional feature selection: evaluation for genomic prediction in man. Sci. Rep. **5**, 10312 (2015). https://doi.org/10.1038/srep10312

9. Tsai, F.S.: Dimensionality reduction for computer facial animation. Expert Syst. Appl. **39**(5), 4965–4971 (2012). https://doi.org/10.1016/j.eswa.2011.10.018

10. Guyon, I., Elisseeff, A.: An introduction to variable and feature selection. J. Mach. Learn. Res. **3**, 1157–1182 (2003)

11. Lewis, P.: The characteristic selection problem in recognition systems. IRE Trans. Inf. Theory **8**(2), 171–178 (1962)

12. Chien, Y., Fu, K.S.: On the generalized Karhunen-Loève expansion (Corresp.). IEEE Trans. Inf. Theory **13**(3), 518–520 (1967)

13. Weston, J., Mukherjee, S., Chapelle, O., Pontil, M., Poggio, T., Vapnik, V.: Feature selection for SVMs. In: Advances in Neural Information Processing Systems 13, pp. 668–674. MIT Press (2000)

14. Kozachenko, L., Leonenko, N.N.: Sample estimate of the entropy of a random vector. Problemy Peredachi Informatsii **23**(2), 9–16 (1987)

15. Fisher, R.A.: XV-the correlation between relatives on the supposition of mendelian inheritance. Earth Environ. Sci. Trans. R. Soc. Edinb. **52**(2), 399–433 (1919)

16. Deb, K., Pratap, A., Agarwal, S., Meyarivan, T.: A fast and elitist multiobjective genetic algorithm: NSGA-II. IEEE Trans. Evol. Comput. **6**(2), 182–197 (2002)

17. Heiman, G.W.: Understanding Research Methods and Statistics: An Integrated Introduction for Psychology. Mifflin and Company, Houghton (2001)

18. Cox, D.R.: The regression analysis of binary sequences. J. Roy. Stat. Soc. Ser. B (Methodol.) **20**(2), 215–232 (1958)

19. Guyon, I., Weston, J., Barnhill, S., Vapnik, V.: Gene selection for cancer classification using support vector machines. Mach. Learn. **46**(1–3), 389–422 (2002)

20. Welch, B.L.: The generalization of student's problem when several different population variances are involved. Biometrika **34**(1/2), 28–35 (1947)

21. Krzywinski, M., Altman, N.: Points of significance: comparing samples-part I. Nat. Methods **11**(3), 215 (2014)

22. Pedregosa, F., et al.: Scikit-learn: machine learning in Python. J. Mach. Learn. Res. **12**, 2825–2830 (2011)

23. Casalicchio, G., et al.: OpenML: an R package to connect to the machine learning platform OpenML. Comput. Statistics **34**(3), 977–991 (2017). https://doi.org/10.1007/s00180-017-0742-2

24. Garrett, A.: inspyred (version 1.0.1) inspired intelligence (2012). https://github.com/aarongarrett/inspyred

25. Vanschoren, J., van Rijn, J.N., Bischl, B., Torgo, L.: OpenML: networked science in machine learning. SIGKDD Explor. **15**(2), 49–60 (2013). https://doi.org/10.1145/2641190.2641198

26. Dua, D., Graff, C.: UCI machine learning repository (2017). http://archive.ics.uci.edu/ml

27. Quinlan, J.R.: Simplifying decision trees. Int. J. Man Mach. Stud. **27**(3), 221–234 (1987)

28. Siebert, J.P.: Vehicle recognition using rule based methods (1987)
29. Guyon, I., Gunn, S., Ben-Hur, A., Dror, G.: Result analysis of the NIPS 2003 feature selection challenge. In: Advances in Neural Information Processing Systems, pp. 545–552 (2005)
30. Guyon, I.: Design of experiments of the NIPS 2003 variable selection benchmark. In: NIPS 2003 Workshop on Feature Extraction and Feature Selection (2003)

Fast Evolutionary Algorithm for Solving Large-Scale Multi-objective Problems

Anna Ouskova Leonteva[(⊠)], Pierre Parrend, Anne Jeannin-Girardon,
and Pierre Collet

ICube CSTB, 67000 Strasbourg, France
anna.ouskova-leonteva@etu.unistra.fr,
{pierre.parrend,anne.jeannin,pierre.collet}@unistra.fr

Abstract. This paper proposes a fast evolutionary algorithm for large-scale multi-objective optimization problems (MOPs), which widely exist in real-world applications [3,6]. Many well-established multi-objective evolutionary algorithms (MOEAs) can not ensure necessary Runtime (RT) and values of performance metrics (Hypervolume (HV), Inverted Generational Distance (IGD)) for such kind of MOPs. The proposed archive-based algorithm provides better values of mentioned metrics due to its low complexity, simplified architecture and efficiency of genetic operators. Experimental results on three-objective and on two-objective benchmark suites (DTLZ [15], COCO 2018 Blackbox Optimization Benchmark (BBOB-biobj) [8]) demonstrate superiority of suggested algorithm in terms of performance metrics values and of RT over referenced MOEAs.

Keywords: Large-scale multi-objective optimization problems ·
Genetic operators · Adaptation · Performance · Runtime

1 Introduction

With growing interest in the high-fidelity computer simulations [17], the necessity in algorithms to solve large-scale MOPs is increasing. In order to optimize MOPs in high-dimensional continuous search spaces, a large population size needs to be used. Despite of a huge variety of MOEAs, most of them cannot be directly implemented to solve such problems because of their high computational complexity. Moreover, a low efficiency of existing genetic operators does not allow to achieve necessary values of performance metrics. A choice of parameter values for genetic operators significantly influences on the efficiency of MOEAs [1] and requires a comprehensive knowledge about algorithms and MOPs. Taking into account that the situation changes per problem instance and during evolution, to find one static parameter configuration for optimal performance is still a hard issue. This paper proposes a fast MOEA, called FastEMO, which uses archive-based approach of ASREA [7]. In order to improve performance metric values of ASREA, the following contributions were made:

© Springer Nature Switzerland AG 2020
L. Idoumghar et al. (Eds.): EA 2019, LNCS 12052, pp. 82–95, 2020.
https://doi.org/10.1007/978-3-030-45715-0_7

- Update Archive Method (UAM) with selection of parent population, which allows to simplify the architecture of algorithm and to improve performance metrics;
- new adaptive crossover operator, which ensures continuously-increasing accuracy of HV during the process of evolution;
- implementation an additional function of mutation step size estimation, depending on the input data range in self-adaptive Gaussian mutation operator, for ensuring a rapid convergence rate in short RT.

By these improvements, FastEMO provides a high performance on different MOPs. According to experimental results on 3-objective DTLZ benchmarks suite, FastEMO has advantages over ASREA and 4 well-established algorithms (NSGA3, IBEA, MOEA/D, CDAS) in values of performance metrics and in CPU RT for the maximum budget $= 500000 \times D$ function evaluations (where search space dimension $D = 10$). Also FastEMO shows better RT in number of evaluations in the initial and final stages of optimization versus the best 2016 portfolio of COCO on separable, moderate and ill-conditions BBOB-biobj tests (55 functions) on $D = 40$.

FastEMO is integrated in the open-source platform EASEA (EAsy Specification of Evolutionary Algorithms), which is publicly available and designed to simplify utilization of genetic and evolutionary algorithms.

This paper is organized as follows: Sect. 2 details the proposed approach. In Sect. 3, comparative experimental results on different benchmarks suites are presented. Finally, the conclusions are drawn in Sect. 4.

2 Proposed Approach

2.1 Archive-Based Algorithm

The short description of the proposed algorithm is the following:

- as is usual in evolutionary algorithms, initial population (with size N) is created, using random values;
- initial population is evaluated and copied to parent population, which size is equal N at the first generation;
- two parents are selected from the archive, using a binary tournament selection based on the Pareto domination rules. If both solutions are incomparable, the parent with larger crowding distance value is selected. At the first generation, parents are selected from initial population due to the archive is empty;
- parents are crossed by new adaptive crossover (NAC) operator (see Sect. 2.3);
- new offspring mutates by modified self adaptive Gaussian operator (see Sect. 2.4);
- offspring population is evaluated;
- update archive method (UAM) with selection of next parent population is applied in order to create a new parent population (see Sect. 2.2).

The pseudo-code of FastEMO is stated in Algorithm 1:

Result: *result population*
Generate and Evaluate *init population*
Copy *init population* to *parent population*
while *stopping criterion is not met* **do**
| **while** *offspring population is not full* **do**
| | Select both parents from *parent population* by binary tournament
| | selection
| | Apply NAC on selected parents (see section 2.3)
| | Apply Self-Adaptive Gaussian Mutation on offspring (see section 2.4)
| | Evaluate offspring
| | Add offspring to *offspring population*
| **end**
| UAM *archive population* (see Algorithm 2)
end
Copy solutions from *archive population* to *result population*

Algorithm 1: FastEMO pseudo code

2.2   UAM with Selection of Next Parent Population

As ASREA, FastEMO has low computational complexity $O(man)$ (where m is the number of objectives, a is the size of archive, n is the population size) due to computation of small Pareto Front, limited to the archive size. The optimal archive size was determined experimentally. After many tests, it was found, that with the size of archive is equal $15 \times m$, the best values of performance metrics were achieved. Unlike ASREA, FastEMO does not use a procedure of stochastic ranking for offspring population and of selection strategy for next parent population. Instead, in FastEMO, a much simpler architecture is proposed, which ensures higher HV and IGD precision in the same CPU RT. This is done thanks to the following algorithm:

- only non-dominated solutions are retained to archive;
- individual with the smallest value of crowding distance is deleted from archive in case of the archive overflow;
- new parent population size is reduced to archive size, after the first generation;
- all individuals from archive population are copied to the new parent population;
- the size of archive is increased to initial population size on last generation.

The pseudo-code of UAM is shown in Algorithm 2.

Result: next parent population, update archive population
if *last generation* **then**
  | Resize *archive population* from A to maximum
end
while *Index of Offspring < Offspring Population Size* **do**
  Take a New Offspring
  Set Index of Current Archive Individual = 0
  while *Index of Archive Individual < Archive Size* **do**
    if *New Offspring is dominated by a Current Archive Individual* **then**
      | reject New Offspring and go to next one from offspring population
    else
      if *Current Archive Individual is dominated by New Offspring* **then**
        | remove dominated individual from archive
      else
        | increment index of Current Archive Individual : go to next
        | individual in archive
      end
    end
  end
  Add the New Offspring into archive;
  if *Archive Size = Maximal Archive Size + 1* **then**
    if *last generation = true* **then**
      | return
    end
    Find the worst individual in archive by crowding distance
    Delete the worst individual from archive
  end
end
if *first generation* **then**
  | Resize *parent population* from size *init population* to size of archive
end
Replacement *parent population* by individuals from *archive population*

Algorithm 2: Update Archive Method

The proposed method allows to improve performance metrics due to increasing number of obtained non-dominated solutions and more efficient architecture.

2.3 New Adaptive Crossover Operator

In this section, we introduce a new adaptive crossover. Figure 1 illustrates its structure scheme. The suggested operator works in 2 steps:

– on the first step, an offspring localization range is determined according to
BLX-α crossover [16] rule, by which a possible offspring value depends on
the difference between parent solutions. Instead of the static parameter α,
NAC uses adaptive parameter of localization, which depends on the number
of current generation and is described in Eq. 1:

$$\alpha = P_{min} \times \left(\frac{P_{max}}{P_{min}}\right)^{\left(\frac{gen_i}{gen_{max}}\right)} \tag{1}$$

where P_{min} and P_{max} - minimal and maximal value of crossover probability
(we recommend to choose $P_{min} = 0.01$ and $P_{max} = 0.99$), gen_i - number of
current generation, gen_{max} - total number of generations, This dependence of
α value from the number of current generation was obtained experimentally.
The changing curve of α from the number of current generation is shown
in Fig. 2(a). Minimum and maximum values of localization are determined
according to Eqs. 2 and 3:
if $p_1^i < p_2^i$

$$min = p_1^i - |(p_1^i - p_2^i)| \times \alpha, \quad max = p_2^i + |(p_1^i - p_2^i)| \times (1 - \alpha) \tag{2}$$

otherwise,

$$min = p_2^i - |(p_1^i - p_2^i)| \times (1 - \alpha), \quad max = p_1^i + |(p_1^i - p_2^i)| \times \alpha \tag{3}$$

where p_1^i and p_2^i - a gene i in the parent 1 and parent 2, α - the adaptive
parameter, which is going from Eq. 1. At the end of the first step, a preliminary
offspring determination is carried out by random selection from the obtained
localization range.
– on the second step, a new offspring is obtained by the rule of Arithmetic
crossover. But instead a static crossover parameter, an adaptive parameter β
is used. It is described in Eq. 4.

$$\beta = P_{max} \times \left(\frac{P_{min}}{P_{max}}\right)^{\left(\frac{gen_i}{gen_{max}}\right)^2} \tag{4}$$

where P_{min}, P_{max}, gen_i and gen_{max} are the same as in Eq. 1. This equation
was also obtained experimentally. The changing curve of β from the number
of current generation is shown in Fig. 2(b). The final offspring is determined
in Eq. 5 as a value between obtained offspring on the first step and parent
value, taking into account parameter β:

$$o'^i_1 = o_1^i \times (1 - \beta) + p_1^i \times \beta \tag{5}$$

where o'^i_1 - a new offspring gene, o_1^i - a offspring gene from the first step, p_1^i
and p_2^i are the same as in (2) and (3). Applying this two-step constructed
crossover with two adaptive parameters allows to obtain the continuously-
increasing accuracy of HV.

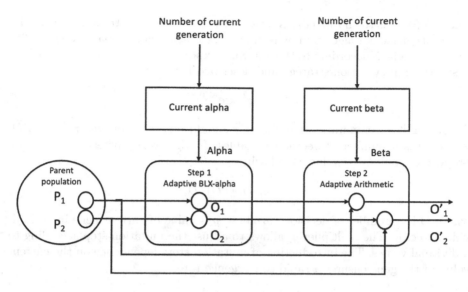

Fig. 1. New adaptive crossover operator

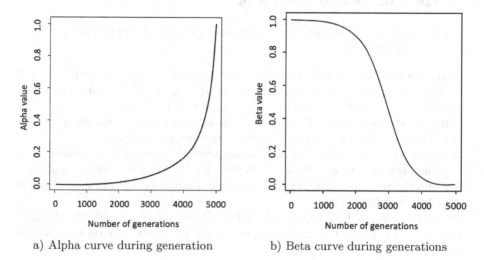

a) Alpha curve during generation   b) Beta curve during generations

Fig. 2. Adaptive coefficients

2.4   Modified Self-adaptive Gaussian Mutation

The Gaussian distribution was chosen as a basis of the mutation operator, because it covers the whole solution space, it does not show drift and it can scale the randomly drawn samples in the whole solution space. The Gaussian mutation operator adds noise to each gene g_i of solution vector as follow:

$$g_i^{t+1} = g_i^t + N(0, \sigma), \qquad (6)$$

where $N(0, \sigma)$ is the normal distribution with zero mean and the standard deviation σ. To make our algorithm more efficient, the self-adaptive parameter modification is applied, according to the following exponential one-step method, which is standard in evolution strategy and described here [2]:

$$\sigma^{t+1} = \sigma^t \times e^{N(0,\tau)}, \tag{7}$$

According [2], the mutation parameter τ should be set as follows: $\tau = 1/2\sqrt{D}$, where D is a number of decision variables. In this work, instead of Eq. 7, we propose to compute τ value as following:

$$\tau = g_i * log(D)/D, \tag{8}$$

where D is a number of decision variables, g_i - value of current gene in an individual vector. The multiplier g_i allows to adjust the mutation step according to individual values. The introduction of mutation step dependence on the current value of the gene ensures a rapid convergence rate.

3   Experiments and Validation

In order to objectively evaluate the performance of FastEMO the following experiments were carried out:

- comparison of NAC against BLX-α and Arithmetic operators on different 2-objective functions with a search space dimension D = 40, for proving a positive influence of NAC on the HV accuracy of FastEMO;
- comparison of modified self-adaptive Gaussian mutation against simple Gaussian mutation for checking its advantage in convergence rate;
- comparison of FastEMO against archive-based ASREA and against four reliable-performing algorithms on 3-objective DTLZ test suite for confirming effectiveness of UAM, NAC and modified mutation operator on classical MOPs in the 10D space;
- comparison of FastEMO against best BBOB-2016 results on 55 BBOB-biobj functions (for D = 40) in order to check FastEMO advantage on large-scaled MOPs.

3.1   Experimental Setup

In order to estimate FastEMO performance on 55 BBOB-biobj functions (on 5–10 instances) of COCO (Comparing Continuous Optimizer Platform) [8] the following method is used: the central performance measure is the RT in terms of the number of evaluations conducted on a given problem until a given target value is hit (HV precision). The objectives of experiments are to achieve the first HV precision ($1e+0$) and the highest HV precision ($>= 1e-05$) in the smallest number of evaluations. The initial population is sampled uniformly at random in $[0, 1]^D$.

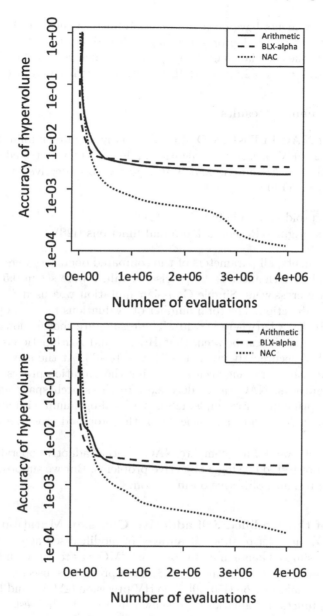

Fig. 3. HV accuracy profiles obtained by FastEMO with different crossovers on f2, f28 bi-bbob test functions

For other benchmarks suites, a total CPU time execution for given number of generations and a mean value of RT per one generation is used. The objective is to achieve the highest value of HV and the lowest value of IGD in the shortest CPU RT. Evaluation of the test results was also carried out by analysing graphs of Pareto fronts.

All the experiments have been run on an Intel(R) Pentium(R) CPU 4405U @ 2.10 GHz 4 processors laptop on a single thread via the platform EASEA version 2, using the code language (C++) and compiler (g++ 5.4.0). We use the platform EASEA version 2 for all MOEAs without further parameter tuning.

3.2 Experimental Results

Efficiency of NAC in FastEMO. In order to investigate an influence of NAC operator on the HV accuracy of algorithm, FastEMO was tested with three crossovers (BLX-α, Arithmetic and NAC) on two bi-objective functions from the platform COCO [9]:

- Sphere/Ellipsoid separable functions (f2),
- Rosenbrock original/Rosenbrock original functions (f28).

For fair experiments, all parameters of the compared operators were tuned for a relatively good performance. To be precise, α values were set in 0.5 for BLX-α and Arithmetic crossovers. Simple Gaussian mutation was used for all experiments in this subsection. The total number of evaluations for each optimization problem = 4000000 × D (where D = 40). The results are shown shown in Fig. 3.

From the Fig. 3, it can be seen, that BLX-α and Arithmetic crossovers get stuck at the HV accuracy between 1e−02 and 1e−03. At the same time, NAC shows continuously-increasing accuracy during the evolution process due to two adaptive parameters. NAC can widely explore the search space and helps to avoid a local optimum (for example, f28 has a local optimum with an attraction volume of about 25%). So, the efficiency of the proposed crossover operator is confirmed.

In this article, we did not compare NAC to other adaptive operators, as well as, NAC was not tested on single-objective problems. But we suppose that NAC is not specific to multi-objective optimization.

Efficiency of the Modified Self-adaptive Gaussian Mutation Operator in FastEMO. In order to show advantages of modified self-adaptive Gaussian mutation over simple Gaussian mutation, FastEMO was tested with both operators on two bi-objective functions (f2, f28) from previous subsection and on one highly multi-modal function (f50) [9]. The HV precision (ΔHV) and RT in terms of number of function evaluations are shown in Table 1 (the best values are in bold face) for total evaluations budget = 4000000 × D functions. It can be seen from Table 1, that FastEMO with modified self-adaptive mutation outperforms the results obtained with simple Gaussian mutation on separable-separable and moderate-moderate test functions in RT and HV precision. But both mutation operators are not efficient in case of multi-modal function.

Table 1. Comparison modified self-adaptive Gaussian mutations against simple Gaussian mutation on f2, f28, f50 bi-bbob functions

Function	Mutation	RT	ΔHV
f2	Simple	4000000	1e−03
	Self-adaptive	**1414000**	**1e−08**
f28	Simple	4000000	1e−03
	Self-adaptive	**3739000**	**1e−08**
f50	Simple	4000000	1e+0
	Self-adaptive	4000000	1e+0

Comparison FastEMO Against ASREA, MOEA/D, NSGA3, CDAS and IBEA. We now compare FastEMO to the basic method ASREA, as well as with four well-established algorithms: MOEA/D [10], NSGA3 [11], CDAS [13], IBEA [12] (with adaptive epsilon-indicator) on 3-objective DTLZ test functions for total large budget of $500000 \times D$ (where $D = 10$) function evaluations. Each

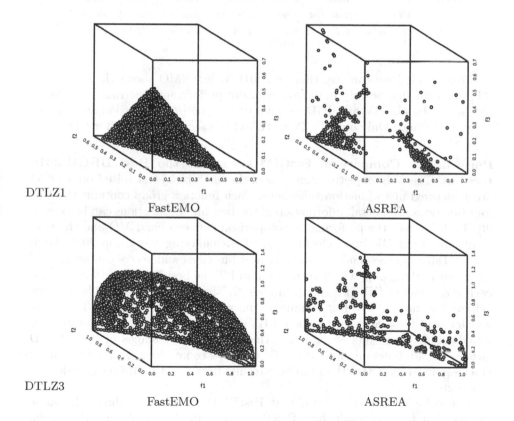

Fig. 4. Pareto fronts of FasteEMO and ASREA on DTLZ1 and DTLZ3 benchmark

algorithm was run 4 times. Experimental results are presented in Fig. 4 and Table 2 (the best values are in bold face). Figure 4 shows Pareto fronts obtained by ASREA and FastEMO on DTLZ1 and DTLZ3 functions and proves the advantage of the new improvements.

Table 2. Results comparison on DTLZ tests

Problem	Metrics	FastEMO	ASREA	MOEA-D	NSGA3	IBEA	CDAS
DTLZ1	HV	**0.823**	0.735	0.806	0.803	0.810	0.807
	IGD	**7.9e−05**	1.1e−03	2.0e−04	2.7e−04	2.2e−04	1.9e−04
	CPU RT(s)	**4.1**	**4.1**	11.5	15.2	123.2	13.2
DTLZ3	HV	**0.467**	0.308	O.443	0.426	0	0.443
	IGD	**1.6e−04**	3.2e−03	3.7e−04	6.2e−04	2.46	4.0e−04
	CPU RT(s)	**5.5**	**5.5**	12.1	14.09	116.3	13.2
DTLZ5	HV	0.095	0.089	0.095	0.095	0.095	**0.096**
	IGD	5.6e−06	5.3e−05	8.3e−06	8.1e−06	3.5e−05	**2.0e−06**
	CPU RT(s)	**3.8**	**3.8**	11.2	18.4	126.1	17.8
DTLZ7	HV	**0.333**	0.321	0.294	0.318	0.324	0.325
	IGD	**3.0e−04**	8.5e−04	1.9e−03	9.0e−04	2.3e−03	6.0e−04
	CPU RT(s)	**6.2**	**6.2**	11.4	25.4	122.14	22.6

From Table 2 we can see, that as ASREA, FastEMO shows the best CPU RT, but with significantly improved values of performance metrics. It achieved the best values of HV and IGD on majority of the problems compared to the other algorithms. Only on DTLZ5 FastEMO has second results after CDAS.

Performance Comparison FastEMO Against the Best BBOB-2016 Results. The 55 bi-objective functions of the bbob-biobj test suite from COCO are structured in 15 function subgroups. Each function group contains three or four functions. The detail information of the benchmark functions can be seen in [9]. Table 3 shows the performance comparison between FastEMO and the best results from the Blackbox Optimization Benchmarking Workshop 2016. Each row in Table 3 corresponds to one group of functions and shows the number of functions in the group, which outperform in RT the best BBOB-2016 results for each target values (HV precision). Simbol "X" means that both results did not achieve the target. The last column is the number of functions in the group, which reached the last target value (Δ HV = 1e−06) versus the best results from BBOB-2016. The results demonstrate, that for total large budget of 4000000 × D (where D = 40) function evaluations, FastEMO is competitive with, and in many cases significantly better than the best results from BBOB-2016 on bi-objective functions.

It can be seen from Table 3, that FastEMO is faster (or shows the same results) than the best results from BBOB-2016 in initial stage of the optimization ($\Delta HV = 1e+0$) on 83% of functions.

Table 3. Summary of performance comparisons between FastEMO and the best results from the Blackbox Optimization Benchmarking Workshop 2016

Group (Total number of functions in group)	Δ HV				Success
	1e+0	1e−02	1e−03	1e−05	
separable - separable (3)	2	2	3	2	2 vs 0
separable - moderate (4)	2	1	4	4	4 vs 0
separable - ill-conditioned (4)	4	2	4	3	3 vs 0
separable - multi-modal (4)	4	1	1	1	1 vs 0
separable - weakly-structured (4)	4	2	2	3	3 vs 0
moderate - moderate (3)	3	2	2	3	3 vs 1
moderate - ill-conditioned (4)	4	1	4	3	3 vs 0
moderate - multi-modal (4)	3	2	2	X	0 vs 0
moderate - weakly-structured (4)	4	3	3	4	4 vs 0
ill-conditioned - ill-conditioned (3)	3	2	3	2	2 vs 0
ill-conditioned - multi-modal (4)	4	3	X	X	0 vs 0
ill-conditioned - weakly-structured (4)	4	1	1	2	2 vs 0
multi-modal - multi-modal (3)	2	0	X	X	0 vs 0
multi-modal - weakly structured (4)	2	1	1	1	1 vs 0
weakly structured - weakly structured (3)	1	0	0	0	0 vs 1

In the middle stage ($\Delta HV = 1e-2$) of the optimization, FastEMO is slightly slower than the best BBOB-2016 results and it shows better performance only on 46% of problems. Specially, FastEMO demonstrates worse results on multi-modal and weakly structured functions. But the next precision $\Delta HV = 1e-3$ FastEMO achieved faster than the referenced results on 54% of functions.

In the last stage, FastEMO solved with success (Δ HV = 1e−06) 56% of functions (28 over 55 functions). Whereas the best BBOB-2016 results demonstrate only 2 over 55 functions, that have been completed with the same accuracy. This fact confirms, that FastEMO performs a high HV accuracy within a small number of evaluations in the 40D space.

On multi-modal functions, both algorithms are dropped into local minimum and have shown relatively close results. But on multi-modal - multi-modal group functions the best BBOB-2016 results demonstrate a slightly better accuracy $\Delta HV = 1e-02$ than FastEMO $\Delta HV = 1e-01$ due to a larger computational budget.

Experiments have been done on 5 instances (from 5 to 10) [9] of BBOB-2018. It was found out, that an impact of each instance on the results is not significant for FastEMO.

4 Conclusion

We introduced and evaluated a new variant of archive-based multi-objective optimization algorithm, called FastEMO. As ASREA [7], it has low computational complexity (O(man)) due to the use of external archive. The proposed approach differs from ASREA in that it applies the new update archive method (UAM), the new adaptive crossover (NAC) and modified self-adaptive Gaussian mutation, that helps to handle high-dimensional continuous problems with better accuracy in short RT. The new algorithm was compared to ASREA, NSGA3, MOEA/D, CDAS and IBEA on 3-objective DTLZ functions, in order to reveal the benefits of new improvements. According to results in Table 2, FastEMO appears to be robust to work with total large budget of $500000 \times D$ function evaluations and demonstrates better values of performance metrics and CPU RT. Then FastEMO was extensively compared with the best results from the Blackbox Optimization Benchmarking Workshop 2016 on 55 bi-objective test functions from COCO 2018 BBOB. From experimental results, shown in Table 3 on the 40D space, FastEMO caught up in performance and overtook the best results from the Blackbox Optimization Benchmarking Workshop 2016 in many cases on the separable, moderate and ill-conditions functions. In further work an additional performance optimization for resolving multi-modal and weakly structured MOPs might be done.

References

1. Lobo, F.J., Lima, C.F., Michalewicz, Z. (eds.): Parameter Setting in Evolutionary Algorithms, vol. 54. Springer, Heidelberg (2007). https://doi.org/10.1007/978-3-540-69432-8
2. Back, T.: Self-adaptation in genetic algorithms. In: Proceedings of the First European Conference on Artificial Life, pp. 263–271. MIT Press, Cambridge (1992)
3. Mei, F., Cao, Q., Jiang, H., Tian, L.: LSM-tree managed storage for large-scale key-value store. IEEE Trans. Parallel Distrib. Syst. **30**(2), 400–414 (2019)
4. Kirkpatrick, S., Gelatt, C.D., Vecchi, M.P.: Optimization by simulated annealing. Science **220**(4598), 671–680 (1983)
5. Qin, A.K., Suganthan, P.N.: Self-adaptive differential evolution algorithm for numerical optimization. In: 2005 IEEE Congress on Evolutionary Computation, vol. 2, pp. 1785–1791. IEEE (2005)
6. Goh, S.K., Tan, K.C., Al-Mamun, A., Abbass, H.A.: Evolutionary big optimization (BigOpt) of signals. In: 2015 IEEE Congress on Evolutionary Computation (CEC), pp. 3332–3339. IEEE, May 2015
7. Sharma, D., Collet, P.: An archived-based stochastic ranking evolutionary algorithm (ASREA) for multi-objective optimization. In: Proceedings of the 12th Annual Conference on Genetic and Evolutionary Computation, pp. 479–486. ACM (2010)
8. COCO: Comparing Continuous Optimizers (2018). https://github.com/numbbo/coco
9. Tusar, T., Brockhoff, D., Hansen, N., Auger, A.: COCO: the bi-objective black box optimization benchmarking (bbob-biobj) test suite (2016)

10. Zhang, Q., Li, H.: MOEA/D: a multiobjective evolutionary algorithm based on decomposition. IEEE Trans. Evol. Comput. **11**(6), 712–731 (2007)
11. Deb, K., Jain, H.: An evolutionary many-objective optimization algorithm using reference-point-based nondominated sorting approach, part I: solving problems with box constraints. IEEE Trans. Evol. Comput. **18**(4), 577–601 (2013)
12. Zitzler, E., Künzli, S.: Indicator-based selection in multiobjective search. In: Yao, X., et al. (eds.) PPSN 2004. LNCS, vol. 3242, pp. 832–842. Springer, Heidelberg (2004). https://doi.org/10.1007/978-3-540-30217-9_84
13. Sato, H., Aguirre, H.E., Tanaka, K.: Controlling dominance area of solutions and its impact on the performance of MOEAs. In: Obayashi, S., Deb, K., Poloni, C., Hiroyasu, T., Murata, T. (eds.) EMO 2007. LNCS, vol. 4403, pp. 5–20. Springer, Heidelberg (2007). https://doi.org/10.1007/978-3-540-70928-2_5
14. Hansen, N., Auger, A., Brockhoff, D., Tušar, D., Tušar, T.: COCO: performance assessment. arXiv preprint arXiv:1605.03560 (2016)
15. Deb, K., Thiele, L., Laumanns, M., Zitzler, E.: Scalable multi-objective optimization test problems. In: Proceedings of the 2002 Congress on Evolutionary Computation, CEC 2002 (Cat. No. 02TH8600), vol. 1, pp. 825–830. IEEE (2002)
16. Herrera, F., Lozano, M., Sánchez, A.M.: A taxonomy for the crossover operator for real-coded genetic algorithms: an experimental study. Int. J. Intell. Syst. **18**(3), 309–338 (2003)
17. Simpson, T.W., Booker, A.J., Ghosh, D., Giunta, A.A., Koch, P.N., Yang, R.J.: Approximation methods in multidisciplinary analysis and optimization: a panel discussion. Struct. Multidiscip. Optim. **27**(5), 302–313 (2004)
18. Furqan, M., Hartono, H., Ongko, E., Ikhsan, M.: Performance of arithmetic crossover and heuristic crossover in genetic algorithm based on alpha parameter. IOSR J. Comput. Eng. (IOSR-JCE) **19**(1), 31–36 (2017)

Looking for Energy Efficient
Genetic Algorithms

Francisco Fernández de Vega[1] , Josefa Díaz[1(✉)] , Juan Ángel García[1] ,
Francisco Chávez[2] , and Jorge Alvarado[3]

[1] Computer Architecture Department, University of Extremadura,
C. Santa Teresa de Jornet 38, 06800 Mérida, Spain
{fcofdez,mjdiaz,jangelgm}@unex.es
[2] Computer and Telematics Systems Department, University of Extremadura,
C. Santa Teresa de Jornet 38, 06800 Mérida, Spain
fchavez@unex.es
[3] GEA Group, University of Extremadura,
C. Santa Teresa de Jornet 38, 06800 Mérida, Spain
jorgealvaradodiaz@gmail.com

Abstract. When Evolutionary Algorithms (EAs) are applied to optimization problems, two main measures are taken into account to understand their performance: fitness quality and computing time. These two values are used to compare the performance of different versions of an algorithm, different parameter settings of a single algorithm or even compare a particular EA with other available heuristics. Nevertheless, a new trend in computer science tries to contextualize these features under a new perspective: power consumption. This paper presents a preliminary analysis of the standard genetic algorithm, using two well known benchmark problems, considering their fitness quality, the computing time and also the power consumption when battery-powered devices are used to run them. Results show that some of the main parameters of the algorithm have an impact on instantaneous energy consumption -that departs from the expected behavior, and therefore affects the amount of energy required to run the algorithm. Although we are still far from finding a way to design energy-efficient EAs, we think the results open up a new perspective that will enable us to achieve this goal in the future.

1 Introduction

When evolutionary algorithms are analyzed to understand their performance, researchers firstly consider the quality of solutions obtained, and then, the time required to get such a solution. Although parallel versions have been developed to save computing time, and a plethora of structured models and hardware technologies are available [1], researchers usually take shelter in sequential versions, and only resort to parallel models if the problem they are trying to solve requires days or weeks to be solved.

L. Idoumghar et al. (Eds.): EA 2019, LNCS 12052, pp. 96–109, 2020.
https://doi.org/10.1007/978-3-030-45715-0_8

We will thus focus on this most frequently used approach: the sequential version of the Genetic Algorithm, although the discussion could also be tailored to parallel and distributed versions of the algorithm.

Four decades have passed since GAs were proposed by Holland [2], and over the years this evolutionary based search and optimization heuristic has been run by researchers in any available hardware device that allows them to obtain quality solutions as soon as possible. However, the algorithm power consumption has never been considered as something of interest, although in other computer science areas the subject has already entered the optimization arena [3–6]. Until very recently, the only previous work linking EAs with the optimization of energy consumption, although related to other areas, was presented in [7].

Some review papers have lately included power consumption as one of the issues of interest to be studied when analyzing the behavior of EAs [8], particularly when small battery powered devices are used to run these kind of metaheuristics.

During the last couple of years, we have witnessed the first attempts to study energy-related behaviors in evolutionary algorithms, and the first papers on the subject have already been published [9], that includes some preliminary analysis of power consumption associated to different hardware platforms [10]. This study is particularly relevant when battery powered devices (such as hand-held ones or laptop computers disconnected from the mains) are used, for obvious reasons.

Interest on the topic has emerged, and in [11], for instance, authors analyze the energy consumption behavior when a sequential or parallel genetic algorithm is run. They focus on crossover and mutation operators as well as the fitness function. They point out the need for a deeper study.

Yet, to the best of our knowledge no specific study has been presented that analyzes the impact of the main algorithm parameters, such as population size, on the energy consumed to reach a solution.

This paper thus presents for the first time such an analysis for GAs. Although results are still preliminary, we consider they pave the way to a better understanding of the algorithm under this new perspective, that will allow in the future the design of more energy efficient EAs.

The rest of the paper is organized as follows: in Sect. 2 the motivation for this analysis and a discussion on expected results are presented. Section 3 describes the methodology for the analysis and the benchmark problems selected. Section 4 shows the results obtained, and finally Sect. 5 draws our conclusions and future work.

2 Considering Power Consumption in GAs

As mentioned above, we are interested in analyzing energy footprint of evolutionary algorithms, with particular attention to battery-powered devices that can be used to run them, such as laptop computers, tablets, ipads, mobile phones, all of them with a crucial dependence on energy available in their batteries. We address a first analyses of the relationship between execution time and energy consumption when a classic version of a GA is running on a battery-powered device. We focus on the main parameters of the GAs and their influence on the energy consumption. But before that we must understand why this issue has not been addressed before, and why we think the analysis is worth it.

Let's first consider the standard version of the Genetic Algorithm, when applied to the *One-max* benchmark problem. Among the main parameters for the algorithm and the problem to be solved, we may describe the chromosome size, number of individuals in the population and number of generations, to name just a few.

Thus, the algorithm has to repeat for the number of generations established a series of standard operations: fitness evaluation, selection, crossover and mutation; the larger the population size the higher the number of repetitions of those operations.

From now one, we will focus on the sequential version of the algorithm. This actually means that a single CPU will be devoted to run the algorithm, and the differences on execution time due to Operating System activities can be discarded if data from a large number of runs are used to compute average run times. Actually, the literature has very scarcely considered the influence of Operating System activities on the evolutionary algorithm running time, and probably only when considering parallel models running on a single CPU, given that operating system must manage the scheduling of different processes running simultaneously on a single processor [12].

Therefore, we employ this approach in the theoretical experiment that we describe as our starting point and which in the future will include the operating system in the whole picture of the energy consumption analysis.

2.1   Population Size and Power Consumption

The number of generations, population size and fitness functions are the key components influencing the runtime of a standard Genetic Algorithm. Although certainly different versions of the crossover and mutation operators may take different computing time, we believe that the parameters described above have a greater influence on the time required to run the GA.

If we thus decide to run the algorithm for N generations, the execution will be shorter than when running $N + 1$, and longer than $N - 1$. We do not consider here whether the solution is found before the maximum number of generations is reached, which can be easily ensured by making the problem difficult enough (for instance, by increasing chromosome size in the *one-max* problem). Similarly, if we use I individuals in the population, the runtime will be shorter than when using $I + 1$, and longer than $I - 1$, given that for every individual a fitness evaluation must be computed. Finally, an experiment with a fitness function that needs more time will give rise to a longer execution time.

We must also take into account that when real-life problems are addressed, the perfect solution is typically not found, and the algorithm is configured to stop when a maximum execution time is reached. We take this approach in what follows, fine-tuning the difficulty of the benchmark problems so that an optimal solution is not found for the allotted time.

2.2   Power Consumption and Running Time

Let us consider that the CPU, when running the algorithm, devotes exactly the same effort regardless the specific operation it is performing. If that were the case, this would mean that the instantaneous consumption of energy is the same throughout the experiment, and the total energy consumed could easily be calculated by multiplying the instantaneous consumption of energy (at any given moment throughout the experiment) by the total time required to execute it. Thus, a linear relationship could be described when relating time to power consumption.

This seems to be the case for EA's, which implicitly assumes the importance of computing time, and dispenses with the study of energy consumed, given its direct relationship with computation time.

Assuming the previously described considerations, we could easily build a graph showing the power consumption estimated for different parameter values in the GA. For instance, if a population with size N consumes a given amount of energy during a run, if we expand the population size in a series of experiments, we expect to have power consumption values proportional to the expansion, given that the algorithm will perform a number of additional operations (fitness evaluations, mutations, crossover...) proportional to the new number of individuals in the population. We show the kind of expected behavior in Fig. 1, where every types of line in the graph corresponds to the different population size. Result has been obtained by using values for a single experiment and projecting them to the remaining ones, given that the same behaviour is expected when population sizes are enlarged. We have also considered that the solution is not found along the run. Otherwise, a given experiment would stop and the energy consumed would be different.

The main idea used to build Fig. 1 is that processor's instantaneous energy consumption is constant regardless of the operation performed: the longer the time for its evaluation, the higher the energy required to run it. As we see in the figure, power consumption is proportional to the population size.

2.3   Problem Difficulty and Chromosome Size

A similar analysis could be done for problems where difficulty is related to chromosome size. For instance, when the *one-max* problem is considered, the larger the chromosome the higher the difficulty, and thus the longer the time to be evaluated. Assuming once more that no differences on instantaneous energy consumed by the CPU are present in any of the possible experiments we may launch, the total energy consumed will only depend on time, and a similar graph as that shown in Fig. 1 could be obtained with different chromosome sizes.

We hypothesize that this expected behavior is what has hindered researchers from taking it into account when studying the behavior of the Genetic Algorithm:

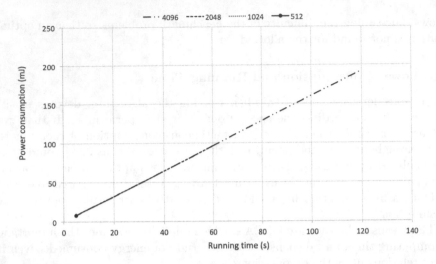

Fig. 1. Expected behavior for the power consumption of GA with different population sizes

if running time and power consumption are proportional values, there is no reason to study both. Once running time is obtained, power consumption can be easily computed.

Nevertheless, it has recently been described that energy is important to decide the more efficient hardware platform to run an algorithm. When efficiency is the key, power consumption is the correct point of view to evaluate the preferred hardware platform [10].

In any case, should we still assume that energy and time are proportional values of a single entity without experimental evidence? Do some of the GA main parameters influence in any way the energy needed to run the algorithm?

We think such an analysis is useful to confirm or discard the assumed hypothesis, and this is what we discuss in the following sections. We only focus on energy consumption in this preliminary analysis, and do not consider quality of solutions found.

3   Methodology

The idea is thus to study if some of the main GA parameters have any influence on the total energy consumed when running the algorithm, or, instead, parameter values have no influence on it.

3.1   Battery Powered Device

As described before, we are mainly interested in energy efficiency when using battery powered devices where energy available is limited, although conclusions could be easily extended to any hardware platform.

In this first study we have chosen a Lenovo Tablet Tab2, A10 - 70F with a Mediatek SoC MT8165, which is a fairly standard representative model among the available options.

This tablet embodies a MediaTek MT8165, 64-bit ARM-based SoC for Android devices, and was launched in 2014. The quad-core processor is manufactured in 28 nm and based on the Cortex-A53 architecture. In addition to the CPU core, it integrates an ARM Mali-760 MP2 GPU and a LPDDR3 memory controller (32-bit, 800 MHz, 6.4 GB/s). It includes four 1.5 Ghz cores.

3.2   Measuring Power Consumption

Once the handheld device has been selected, we need an adequate way to measure the energy consumption of the algorithm. We have chosen an android application that allows us to analyze the actual battery consumption of any Android powered device: PowerTutor [13].

The energy consumption of a given running application may be influenced by the different hardware components that it uses. PowerTutor is a free software tool that allows monitoring the energy consumption of the different hardware components that make up a mobile device. The components that PowerTutor measures are the following: CPU, OLED/LCD, GPS, WIFI, 3G, Audio.

The whole system used to measure energy consumption is composed of the following elements: (i) a web service, and (ii) a plug-in for the PowerTutor app. Next we describe each of the components of the system.

Web Service. The web service implemented incorporates a version of a GA to test the plug-in. The algorithm is run as a web site, within the new implemented plug-in for PowerTutor, so that we can easily obtain the energy it consumes. The algorithm requires two files: a *javascript* file that contains the GA and an *html* file that refers to this *javascript* file.

The GA has been implemented using two different approaches, with and without external libraries. In the first approach, we have coded the GA from scratch. In the second one, external libraries have been used for their coding. We have used NodEO library [14] which includes the necessary functions to create a simple GA in JavaScript using the CommonJS format[1].

Plug-in for PowerTutor. Once the web service is implemented, the algorithm can be run in any Android device and we can measure the energy consumption by means of the PowerTutor plug-in developed.

We have created a plug-in that adds new functionality to PowerTutor. Our plug-in is able to execute a computer task and measure the CPU power consumption resulting from that execution. The plug-in is in charge of monitoring the web environment where the GA is run. The GA must be encoded as a JS file and is executed by the android Webview component.

---

[1] http://requirejs.org/docs/commonjs.html.

Table 1. Parameters & values tested for the *one-max* problem

Population sizes	32, 64, 128, 256,512, 1024 and 2048
Chromosome sizes	32, 64, 128, 256, 512, 1024, ... 32768

Table 2. Parameters & values tested for the *trap* problem, with chromosome size: 200

Population sizes	32, 64, 128, 256, 512 and 1024

3.3 Problems and Parameters Tested

Two well known GA benchmark problems were selected for this preliminary study: *the one-max and the trap function (Deceptive Trap)* problems [15]. The idea is to have different configurations, using some of the main GA parameters when long runs of the experiments are performed, and then compute total energy consumed when different parameters settings are applied. The generational version of the algorithm is run, with a maximum time limit established for the run: 300 s; population size and problem difficulty (chromosome size) were checked for the *one-max* problem, and then, the *trap* function tested to confirm some of the conclusions drawn.

Tables 1 and 2 summarize the experiments performed. For each of the parameter values, 30 independent runs were launched so that we obtained the averaged values shown below. Despite the difficulty established for the problems, some of the runs using large populations were able to find the solution before reaching 300 s. In that cases, mean values were computed for the runs that reached that time step.

4 Results

We began experimenting with the *one-max* problem and testing different chromosome sizes. We must bear in mind that in this specific problem, the larger the chromosome size, the longer the time to find a solution, given that difficulty increases with size. We set the population size to 10, trying to avoid very short running times, thus assuring that we can get energy consumption data for each of the experiment along the 300 s established for the run. Although other possibilities are available, we think this experiment provides relevant information for the goal we pursue.

As we may notice in Fig. 2, being all the curves similar, some differences are present when we tested large chromosome sizes. Moreover, some of the differences seem to deviate from what we might expect when analysing possible anomalies: when using 4096 bits, total energy consumed is larger than when using 16384 bits, although differences are narrow. On the other hand, in order to better appreciate differences when smaller values for this parameter are employed, we show a zoomed version of the lower part of the graph in Fig. 3. Again, some differences may be noticed, although quite narrow again. Yet, the pattern concerning anomalies confirms what was seen previously: sometimes large values for chromosome size consume smaller amount of energy when compared to smaller chromosome sizes.

Therefore, we see that although the behavior considering chromosome size is not very different from the expected one described in the previous section (all of the curves on top of each other), some anomalies are present that deserve further studies in the future.

We decided afterwards to run more experiments using different values for population size (see Table 1), and establishing a large value for the chromosome size (16384 bits) so that we assure the experiments can be run for 300 s.

Results obtained are shown in Fig. 4. We may notice that the figure is completely different from the expected one (as seen in Fig. 1): Instead of all the lines on top of each other, the experiment shows that important differences exist among population sizes regarding energy consumption of the algorithm. Particularly relevant are differences when large population sizes are employed: the energy consumed is smaller when compared to small population sizes, which is a surprising and intriguing result. Although similar anomalies where found for chromosome sizes, differences are much larger when population size is considered.

We thus decided to focus on population sizes for the second problem, the *trap* function, and launched a series of runs using population sizes shown in Table 2. Averaged values over the 30 runs per population size are shown in Fig. 5. A progression of lines are displayed again in Fig. 5, each of them with different energy consumption behavior. An almost perfect gradation of behaviors is displayed: the larger the population size, the greater the energy required to run. Nevertheless, the curve corresponding to population size 512 does not follow the trend, and describes a power consumption behavior quite different from the other ones. Energy consumed with 512 as population size is a 7.9%, 41.23% and 55.53% lower than 64, 128 and 256, respectively. Consequently, population size influences energy consumption, and given its influence on the time to solution, a balance must be found if energy efficiency is looked for. We must report that when large population sizes were used, some of the runs found the solution to the problem before reaching 300 s, so the average values in the last few steps of time includes a smaller number of runs. In any case, this should not affect energy consumption behavior, that differs again from the expected one.

The above described results correspond to data obtained when the GA is running. We have also measured what happens when the tablet is switched on and the algorithm is not launched. Of course, the operating system will have a role, and some devices may be consuming energy, but this is similar to the situation when any program is running, so measuring energy in this situation will help to contextualize previous results.

Therefore, for a proper comparison, what we have done is to remove from the GA -the program running- all of the operations performed, when the program is run but no instruction is performed for one hour. Once more, 30 independent

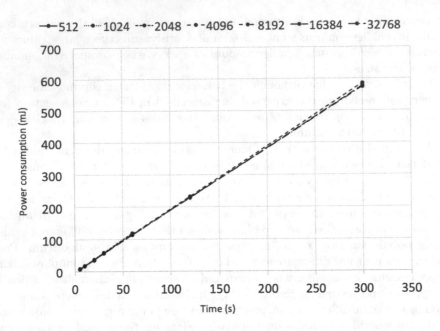

Fig. 2. *One-max* power consumption: analyzing different chromosome sizes. Population size = 10

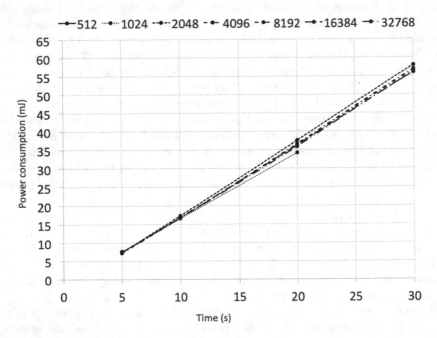

Fig. 3. *One-max* power consumption: analyzing different chromosome sizes (zoom). Population size = 10

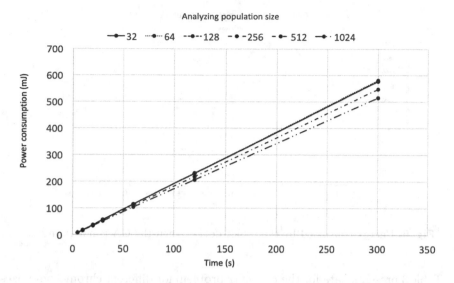

Fig. 4. *One-max* power consumption: analyzing different population sizes (chromosome size = 16384)

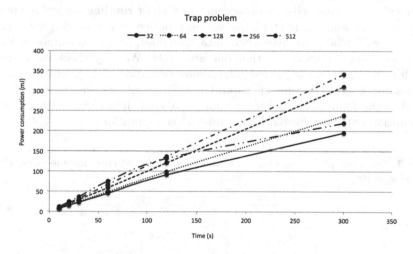

Fig. 5. *Trap* power consumption: analyzing different population sizes.

runs have been tested and values averaged: 30 reset operations on the tablet followed by running the "empty" algorithm and power consumption measured.

Figure 6 shows results. If we focus on timestep 300, the maximum allotted time for GA experiments described above, we see that the total power consumption is less than 300, and this can be compared with results on Fig. 4, that shows values over 500 mJ for *one-max*.

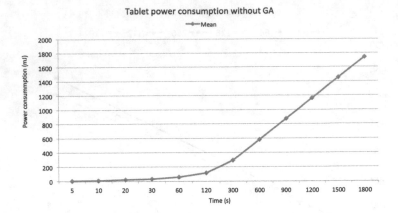

Fig. 6. Energy consumed by the tablet device when the GA is not running.

Table 3 presents data for the *one-max* problem for different chromosome sizes tested with 10 individuals in the population. Each column contains power consumption mean/standard deviation over the 30 runs at different time steps. Empty values in some cells are on account of a short runtime (experiments end before reaching some time step). As we may notice, when we employ a chromosome size of 4096 (and 2048 when data available), energy consumed is larger than when using values above that one after 30 s. We may check for instance the value obtained after 300 s: 585.5 (standard deviation of 4.9), while for larger chromosomes energy consumed is around 577. This confirms what we had already noticed in the figures shown before: that significant differences exist in energy consumption patterns affected by parameters configuration.

Table 3. *One-max* power consumption: analyzing different chromosome sizes. Population size $= 10$

Chromosome size	Time (s)						
	5	10	20	30	60	120	300
512	7.600/0.44						
1024	7.450/0.36	16.532/0.49	34.187/0.61				
2048	7.326/0.53	17.256/0.62	37.528/1.04	57.931/1.51	118.477/1.60		
4096	7.147/0.51	16.676/0.66	36.592/0.94	56.845/1.01	115.755/1.55	232.705/3.27	585.578/4.90
8192	7.456/0.28	16.791/0.63	36.315/0.92	56.435/1.87	113.782/2.02	229.485/2.39	577.026/4.06
16384	7.377/0.36	16.730/0.62	36.307/0.85	55.975/1.57	113.836/1.84	229.275/2.65	576.852/3.73
32768	7.344/0.33	16.678/0.48	35,897/0.78	55.975/1.44	113.472/1.76	228.783/2.66	574.544/7.13
Mean value/Standard deviation	7.386/0.425	16.777/0.623	36.354/1.195	56.632/1.652	115.041/2.562	230.040/3.128	578.256/6.504

On the other hand, Table 4 includes similar information for the *one-max* problem when using different population sizes with a previously established chromosome size: 16384 bits. As we may see for this series of experiments, when large populations are used (512 and 1024), the standard deviation allows us to confidently state that the power consumption is smaller, a behavior that was observed in Fig. 4, and that requires more research in the future.

Table 4. *One-max* power consumption: analyzing different population sizes. Chromosome size = 16384

Population size	Time (s)						
	5	10	20	30	60	120	300
32	7.334/0.36	16.487/0.63	36.056/0.80	55.761/1.06	113.828/1.81	229.618/2.50	576.326/4.79
64	7.295/0.44	16.755/0.78	35.767/1.00	55.994/1.66	113.972/2.19	230.513/3.27	580.024/4.57
128	7.353/0.46	16.791/0.86	36.383/1.41	56.016/1.83	113.738/2.76	229.958/4.29	577.435/8.27
256	7.401/0.42	16.902/0.62	36.089/0.99	55.552/0.90	114.025/1.70	229.713/2.75	576.974/4.56
512	7.415/0.29	16.528/0.57	34.473/0.98	53.178/1.25	108.981/1.69	217.755/2.71	546.043/4.48
1024	7.427/0.23	15.994/0.57	32.953/0.90	51.277/0.99	102.991/1.57	205.073/1.91	513.900/2.95
Mean value/ Standard deviation	7.371/0.37	16.576/0.73	35.287/1.58	54.630/2.22	111.256/4.57	223.772/9.95	561.784/24.96

Table 5 details similar information for the *trap* problem for different population sizes combined with a chromosome size of 200 bits. Results allow again to be confident with differences found. The last row in the three previous tables presents mean values and standard deviation computed over the complete set of tests for each time step established.

All in all, after this series of experiments we can confirm that for the first time we have detected that some of the main parameter values has an effect on GAs power consumption; particularly relevant is population size employed. This could be related to memory usage patterns, that may affect cache access operations. But more experiments are required to fully confirm and understand whether this is a general behavior of EAs, and that a proper parameter configuration may allow to design energy efficient GAs.

Table 5. *Trap* power consumption: analyzing different population sizes. Chromosome size = 200

Population size	Time (s)					
	10	20	30	60	120	300
32	7.63/0.67	14.95/0.63	22.36/0.73	45.14/1.67	91.18/3.37	195.95/65.14
64	8.06/0.66	16.18/0.78	24.45/0.84	48.90/1.14	99.35/3.19	238.89/49.02
128	9.3/1.01	18.74/1.01	28.94/1.34	59.18/1.77	121.50/2.09	310.71/5.46
256	10.47/0.99	21.16/1.22	32.25/1.29	66.73/2.67	136.64/5.59	342.17/42.10
512	11.48/1.01	23.79/1.54	35.96/1.24	73.78/8.20	131.64/40.39	220.00/142.89
Mean value/Standard deviation	10.413/1.668	21.398/3.353	31.645/9.360	105.098/16.990	116.368/27.049	260.954/91.138

5 Conclusions

Although traditionally the quality of fitness and the time required to reach a solution have been the main elements to analyze the behavior of the algorithms, this work proposes and describes the reason to consider energy consumption as a new measure to be applied when analysing GAs behavior.

Few studies have previously included energy consumption as a topic of interest, probably because a linear relationship between computing time and power consumption was assumed.

This paper presents a preliminary analysis on the influence of some of the main GA parameters on its energy consumption patterns. To the best of our knowledge, this analysis is the first to study the impact of GA configuration parameters on the energy required to run the algorithm.

Two well known problems have been selected for the analysis: the *one-max* problem and the *trap* function. Different values for population and chromosome sizes have been tested, and the energy required to run the algorithm for 300 s has been compared with the theoretically expected results -linear relationship between time and energy. In both problems, different anomalies departing from the expected behavior have been found: (i) time and energy behavior does not linearly correlate; (ii) a connection exists among parameter values and power consumption. In any case, anomalies have been found, such as smaller consumption with larger population sizes, that deserves further research if our goal is to find solutions and also reduce energy consumption.

Although experiments have been run in a specific battery-powered android device, we hope to enlarge the study in the future to a larger set of devices, including laptops, Raspberry Pi, etc, so that we can confirm this behavior regardless of the specific underlying hardware, and thus be able to design more energy-efficient evolutionary algorithms in the future.

Acknowledgment. We acknowledge support from Spanish Ministry of Economy and Competitiveness under project TIN2017-85727-C4-{2,4}-P, Regional Government of Extremadura, Department of Commerce and Economy, the European Regional Development Fund, a way to build Europe, under the project IB16035 and Junta de Extremadura, project GR15068.

References

1. Vega, F.F., Pérez, J.I.H., Lanchares, J.: Parallel Architectures and Bioinspired Algorithms, vol. 122. Springer, Heidelberg (2012). https://doi.org/10.1007/978-3-642-28789-3
2. Holland, J.H.: Adaptation in Natural and Artificial Systems: An Introductory Analysis with Applications to Biology. Control and Artificial Intelligence. MIT Press, Cambridge (1992)
3. Albers, S.: Energy-efficient algorithms. ACM Commun. **53**(5), 86–96 (2010)
4. Ye, M., Li, C., Chen, G., Wu, J.: EECS: an energy efficient clustering scheme in wireless sensor networks. In: PCCC 2005, 24th IEEE International Performance, Computing, and Communications Conference 2005, pp. 535–540, April 2005
5. Camilo, T., Carreto, C., Silva, J.S., Boavida, F.: An energy-efficient ant-based routing algorithm for wireless sensor networks. In: Dorigo, M., Gambardella, L.M., Birattari, M., Martinoli, A., Poli, R., Stützle, T. (eds.) ANTS 2006. LNCS, vol. 4150, pp. 49–59. Springer, Heidelberg (2006). https://doi.org/10.1007/11839088_5
6. Heinzelman, W.R., Chandrakasan, A., Balakrishnan, H.: Energy-efficient communication protocol for wireless microsensor networks. In: Proceedings of the 33rd Hawaii International Conference on System Sciences, HICSS 2000, vol. 8, p. 8020. IEEE Computer Society (2000)
7. Gacto, M.J., Alcalá, R., Herrera, F.: A multi-objective evolutionary algorithm for an effective tuning of fuzzy logic controllers in heating, ventilating and air conditioning systems. Appl. Intell. **36**(2), 330–347 (2012)

8. Camacho, D., et al.: From ephemeral computing to deep bioinspired algorithms: new trends and applications. Fut. Gener. Comput. Syst. **88**, 735–746 (2018)
9. Álvarez, J.D., Lao, F.C., Castillo, P., García, J.A., Rodriguez, F., Vega, F.F.: A fuzzy rule-based system to predict energy consumption of genetic programming algorithms. Comput. Sci. Inf. Syst. **15**, 26 (2018)
10. de Vega, F.F., Chávez, F., Díaz, J., García, J.A., Castillo, P.A., Merelo, J.J., Cotta, C.: A cross-platform assessment of energy consumption in evolutionary algorithms. In: Handl, J., Hart, E., Lewis, P.R., López-Ibáñez, M., Ochoa, G., Paechter, B. (eds.) PPSN 2016. LNCS, vol. 9921, pp. 548–557. Springer, Cham (2016). https://doi.org/10.1007/978-3-319-45823-6_51
11. Abdelhafez, A., Alba, E., Luque, G.: A component-based study of energy consumption for sequential and parallel genetic algorithms. J. Supercomput. **75**(10), 6194–6219 (2019). https://doi.org/10.1007/s11227-019-02843-4
12. Fernández, F., Galeano, G., Gómez, J.A.: Comparing synchronous and asynchronous parallel and distributed genetic programming models. In: Foster, J.A., Lutton, E., Miller, J., Ryan, C., Tettamanzi, A. (eds.) EuroGP 2002. LNCS, vol. 2278, pp. 326–335. Springer, Heidelberg (2002). https://doi.org/10.1007/3-540-45984-7_32
13. Yang, Z.: Powertutor-a power monitor for android-based mobile platforms (2012)
14. Guervós, J.M., Castillo, P., Mora, A., Esparcia-Alcázar, A., Santos, V.R.: Nodeo, a multi-paradigm distributed evolutionary algorithm platform in Javascript. In: GECCO 2014 - Companion Publication of the 2014 Genetic and Evolutionary Computation Conference (2014)
15. Deb, K., Goldberg, D.E.: Analyzing deception in trap functions. In: Foundations of Genetic Algorithms, volume 2 of Foundations of Genetic Algorithms, pp. 93–108. Elsevier (1993)

Evolving Fitness Landscapes
with Complementary Fitness Functions

Vincent Hénaux, Adrien Goëffon[✉], and Frédéric Saubion

Université d'Angers (Laboratoire d'Étude et de Recherche en Informatique d'Angers,
LERIA, EA 2645, SFR MathSTIC), Angers, France
{vincent.henaux,adrien.goeffon,frederic.saubion}@univ-angers.fr

Abstract. Given an optimization problem, local search algorithms may
fail to reach optimal solutions when faced to difficult and unsuitable fitness landscapes. Climbing based optimization is sensitive to unexpected
distribution of local optima. In this paper, we aim at modifying the initial fitness landscape of a problem in order to better fit climbing requirements. We propose thus a fitness landscape generation framework based
on an evolutionary process. Preliminary experiments are presented as a
proof of concept.

Keywords: Fitness landscapes · NK landscapes · Local search ·
Function evolution

1 Introduction

Metaheuristics and local search based algorithms are optimization procedures
that guide the search process based on information on the quality of the solutions encountered during the search. In this sense, such techniques can be seen as
an intelligent sampling of the solution space. The quality of solutions is given by
a fitness function which usually corresponds to the objective function of the problem or is closely related to it. Problem solving difficulties appear when the one-
to-one correspondence between solutions and their objective values are difficult
to exploit. Indeed, objective functions leading to multimodal fitness landscapes
reveal what appears to be contradictory information, leading to misled guidance.
Practically speaking, when a search landscape is rugged and has numerous local
optima, the information induced by the objective function cannot be efficiently
exploited by a search algorithm.

The problem targeted here is indeed related to the seminal algorithm selection problem [12] in the context of optimization problems. Given an optimization problem, how to select to best search algorithm according to performance
criteria? This problem has been tackled for many years [2,4,7], leading to efficient algorithm optimization procedures [5,8]. Since many metaheuristics involve
parameters and variable components, various machine learning techniques can
be integrated to search algorithms for tackling difficult optimization problems.

© Springer Nature Switzerland AG 2020
L. Idoumghar et al. (Eds.): EA 2019, LNCS 12052, pp. 110–120, 2020.
https://doi.org/10.1007/978-3-030-45715-0_9

In order to better understand and explain why a given search algorithm may or not efficiently solve a given problem instance, fitness landscape analysis [3] provides relevant high-level information that can be efficiently used in the search process (e.g., detecting specific topological properties of the problem [13, 14]).

The design of an efficient local search algorithm usually focuses on the move strategy, given a natural neighborhood relation and the objective function f_{obj} as fitness function. The move strategy thus aims at overcoming the difficulties inherent in the properties of the landscape defined by the neighborhood relation and the objective function. In this work, we choose to address an alternative view of this algorithm optimization challenge: *Given an instance of a problem and a search algorithm, how to modify the fitness landscape such that the search algorithm becomes more efficient?* Some algorithms may modify the fitness landscape by changing the neighborhood relation [10]. Here, we focus on the fitness function. We propose ways of constructing fitness functions allowing a simple local search algorithm to solve easier a problem instance.

Some problems are said to be easy to solve by a local search process. In this case, the objective function information is assumed to be especially relevant for an iterative search process. The associated fitness landscape is less rugged and contains less local optima. For such problems basic move strategies like hill climbers may be used. However, whatever it be an original multimodal fitness landscape and a climbing strategy, there exist better fitness functions than f_{obj} to reach more efficiently the global optimum, ie. fitness functions whose associated fitness landscapes are easier to climb from random starting solutions—for instance, an optimal one-max like fitness function f_{opt} such that $f_{opt}(x) = -d(x, x_{opt})$, where $d(\cdot, \cdot)$ is a distance and x_{opt} the global optimum. Of course, finding an optimal fitness function is equally difficult that finding the global optimum, but we investigate in this paper to find fitness functions such that derived hill climbers are *better* than a climber guided by means of f_{obj}. We propose here a search process (in the space of fitness functions) which consists to improve fitness functions and thus hill climbers (operating in the original search space). The aim is to determine an alternative climber using an alternative fitness function f_{alt} that globally reaches better solutions (evaluated w.r.t. f_{obj}) than the natural climber based on f_{obj}.

In this paper, we propose two alternative strategies for generating f_{alt}. In a first approach, f_{obj} is accessible during the climbing process and we search for a complementary function f_{comp} such that $f_{alt} = f_{obj} + f_{comp}$. The second approach constitutes the main goal and consists to find a relevant f_{alt} while f_{obj} is considered as a black box function. In both cases, the alternative fitness function is generated by an evolution based process and then compared to similar search procedures based on the initial objective function. In order to experiment our approaches, we focus here on pseudo-Boolean optimization problems based on the NK fitness landscape model [6].

2   Background

As a preliminary, let us recall some basic elements relative to optimization problems, fitness landscapes, NK model and local search.

Definition 1 (Optimization Problem). *A (maximization) optimization problem can be defined by a search space Ω and an objective function f_{obj} : $\Omega \to \mathbb{R}$. An element $s \in \Omega$ will be called a solution and an optimal solution is a solution $s^* \in \Omega$ such that $\forall s \in \Omega, f_{obj}(s) \leq f_{obj}(s^*)$. A pseudo-Boolean optimization problem is such that $\Omega = \{0,1\}^N$.*

The purpose of an optimization algorithm is then to compute optimal solutions from Ω. We focus here on local search based techniques that explore Ω using a neighborhood function $\mathcal{N}$ to move across Ω and a fitness function f to evaluate the solutions (which could be different from f_{obj}). The local search dynamics mainly depend on the structural properties of the triplet $(\Omega, \mathcal{N}, f)$ denoted as *fitness landscape* (Definition 2) and more precisely the compatibility between relations $\mathcal{N}$ and f.

Definition 2 (Fitness Landscape). *A combinatorial fitness landscape is defined by a graph induced by a discrete search space Ω structured with a neighborhood function $\mathcal{N} : \Omega \to 2^{\Omega}$, and a mapping of the graph vertices (solutions) given by a fitness function $f : \Omega \to \mathbb{R}$. A N-dimensional binary fitness landscape is such that $\Omega = \{0,1\}^N$ and $\forall s \in \Omega$, $\mathcal{N}(s) = \{s' \in \Omega \,|\, h(s, s') = 1\}$, $h(\cdot, \cdot)$ referring to the Hamming distance. A solution $s^* \in \Omega$ such that $\forall s \in \mathcal{N}(s^*), f(s) \leq f(s^*)$ is a local optimum.*

Of course, the adaptability of a local search algorithm to an optimization problem also depends on the compatibility between f and f_{obj}. We call search landscape of an optimization problem the derived fitness landscape where $f = f_{obj}$.

The random neighbor NK landscape model [6] can be used express binary fitness landscapes, whose properties are determined by means of two parameters N and K. N is the landscape dimension, and $K < N$ specifies the average number of dependencies per variable and then the *ruggedness* of the landscape.

Definition 3 (NK Landscape). *A NK function $f_{NK} : \{0,1\}^N \to [0,1)$ is defined as $f_{NK}(s) = \sum_{i=1}^{N} \Gamma_i(s_i, s_{l_{i1}}, \ldots, s_{l_{iK}})$, where $l_{ij} \in [\![1, N]\!]$ is the variable index of the j-th variable linked with s_i, and each $\Gamma_i : \{0,1\}^{K+1} \to [0, 1/N)$ is a pseudo-boolean function. A binary fitness landscape can then be described using the NK model with N K-uples (each representing a set of bit interdependencies) and $N \cdot 2^{K+1}$ real or decimal values in [0,1/N). A NK landscape is a binary fitness landscape defined by using a NK fitness function.*

A NK problem instance is an optimization problem expressed with a NK function. In experiments, we will use NK functions as optimization problem instances of tunable search landscape properties. Local search trajectories in

---

Algorithm 1. Climber

    Parameter: a fitness landscape $(\Omega, \mathcal{N}, f)$
    Input: an initial solution $s_0 \in \Omega$
    Output: a local optimum $s^* \in \Omega$
1: $s^* \leftarrow s_0$
2: **repeat**
3:    $I \leftarrow \{s \in \mathcal{N}(s^*) \mid f(s) > f(s^*)\}$
4:    **if** $I \neq \varnothing$ **then**
5:       $s^* \leftarrow \operatorname{select}(I, f)$ // select *is a hill climbing selection heuristic*
6:    **end if**
7: **until** $I = \varnothing$
8: **return** s^*

---

fitness landscapes will be defined by a basic hill climbing algorithm (also denoted as *climber*) as described in Algorithm 1.

Classically, we will also consider iterated local search algorithms (ILS) [9] that consist of alternate a climbing and perturbation processes, which aims at escaping the local optimum reached by the hill climber. Of course, the selection heuristic has an impact on the efficiency of the climber [1] but here we use a fixed simple neighbor strategy to better focus on fitness landscape adaptation.

3   Evolving Alternative Fitness Functions

Given a climber, the purpose of our fitness function evolution process is to improve its adaptability by modifying the initial search landscape of a problem instance $(\Omega, \mathcal{N}, f_{\mathrm{obj}})$. The aim is thus to obtain a new fitness landscape $(\Omega, \mathcal{N}, f)$ that better fits both algorithm and problem instance. We focus on a fixed strategy hill climber (first improvement selection heuristic) whose only variable parameter will be the fitness function used to define a guidance criterion within a very basic neighborhood search. Hence, we aim at generating an appropriate alternative fitness function that modifies the initial objective function f_{obj}, and consequently the fitness landscape. In this paper we experiment and compare two approaches. First, we propose to complement the initial fitness landscape for a basic hill climbing procedure, which lead to modify the set of its local optima and may help a climber to reach, statistically, better solutions (with regards to the initial objective function f_{obj}). Secondly, we use the process used for evolving complementary functions to determine alternative functions from scratch, that is not expressed by means of f_{obj}.

3.1   Complementary Fitness Functions

The complementary fitness function f_{comp} is obtained by evolving a fitness function model with regards to a climber performance criterion. This performance criterion simply evaluates the expected quality (with respect to the objective function f_{obj}) of the local optima (in the sense of the evolved fitness function

$f_{\text{alt}} := f_{\text{obj}} + f_{\text{comp}}$) that can be reached by the local search process from different, randomly selected, starting points. The quality of a fitness function f_{alt} is then given by the expected performance of a hill climbing guided by f, which reflects its ability to solve the original problem.

Therefore, the first expected outcome of such a function learning process is to escape from useless local sub-optima that prevent the algorithm from going on further in its search process (a hill climber necessarily stops at the first local optimum reached). A broader aim is to describe an original mechanism of artificial evolution of local search algorithms by working on fitness functions exclusively.

Algorithm 2 describes the fitness function generation process. Let $f_{\text{obj}} : \Omega \to \mathbb{R}$ be an objective function. Let us fix the search algorithm $\mathcal{A}(f)$ based on fitness function f. As previously stated, $\mathcal{A}$ is here a first improvement hill climbing. We consider that the *quality* of algorithm $\mathcal{A}(f)$ is the expected score (with respect to the objective function f_{obj}) of a local optimum returned by $\mathcal{A}(f)$.

We choose a complementary fitness function model, used as a syntax for the complementary fitness function $f_{\text{comp}} : \Omega \to \mathbb{R}$. For instance, if $\Omega = \mathbb{B}^{256}$, f_{comp} may be expressed by means of a NK instance with $N = 256$ (required) and $K = 1$ (tunable). The mutation operator is then defined with respect to the model (genome type). For instance, considering NK functions, to randomly replace coefficients and links (see Algorithm 3).

The evolution algorithm works as follows. First, a random f_{comp} is generated (in our example, a NK instance), and evaluated. The evaluation estimates the quality of algorithm $\mathcal{A}(f_{\text{obj}} + f_{\text{comp}})$. It consists here to perform $p = 100$ climber runs using fitness function $f_{\text{alt}} := f_{\text{obj}} + f_{\text{comp}}$ and to compute the objective values (w.r.t. f_{obj}) of the 100 reached local optima. Then, f_{comp} is mutated (f'_{comp}) and the quality of the corresponding climber $\mathcal{A}(f_{\text{obj}} + f'_{\text{comp}})$ is statistically compared to $\mathcal{A}(f_{\text{obj}} + f_{\text{comp}})$ thanks to a Wilcoxon rank-sum test. If $\mathcal{A}(f_{\text{obj}} + f'_{\text{comp}}) \succ_{f_{\text{obj}}} \mathcal{A}(f_{\text{obj}} + f_{\text{comp}})$ then f'_{comp} becomes the new current complementary function. This mutation process is repeated until a classic stop criterion is reached.

3.2   Alternative Fitness Functions

Instead of mixing the initial objective function and a complementary function generated thanks to Algorithm 2, we propose to use a fitness function generated from scratch by the same Algorithm 2. The algorithm is directly used considering $f_{\text{alt}} = f_{\text{comp}}$ in the generation process. Our purpose is to assess whether it is preferable to evolve the initial fitness landscape by keeping information from f_{obj} or to build a different landscape from scratch. Note that our purpose appears very different from surrogate model generation since the quality of the generated landscape is evaluated with regards to the climber ability to perform interesting runs, in an expected smoother landscape. Of course reached local optima are expected to coincide with the better optima of f_{obj}, but the resulting fitness model is not expected to share other properties with the initial fitness landscape.

Algorithm 2. Fitness function evolution

Parameters: a fitness function model ($\mathcal{M}$), a mutation parameter (ζ), a solution space (Ω), a neighborhood function ($\mathcal{N}$), a statistical test ($\mathcal{T}$), a statistical significance threshold (z_{min}), the number of hill climbing runs for evaluate fitness functions (p), the maximum number of consecutive non-improving mutations (maxmut).

Input: an objective function f_{obj}.

Output: an alternative fitness function f_{alt}.

1: $f_{comp} \leftarrow$ initialize$_{\mathcal{M}}()$
2: **repeat**
3:      **for** $j \leftarrow 1$ to p **do**
4:          randomly select $s \in \Omega$
5:          $S_j \leftarrow$ hill_climbing$_{(\Omega,\mathcal{N},f_{obj}+f_{comp})}(s)$ // S is a vector of local optima w.r.t. $f_{alt} := f_{obj} + f_{comp}$
6:          $F_j \leftarrow f_{obj}(S_j)$ // F is a vector of scores
7:      **end for**
8:      nbmut $\leftarrow 0$
9:      **repeat**
10:         improve $\leftarrow$ false
11:         $f'_{comp} \leftarrow$ mutate$_{\mathcal{M},\zeta}(f_{comp})$
12:         **for** $j \leftarrow 1$ to p **do**
13:             randomly select $s \in \Omega$
14:             $S'_j \leftarrow$ hill_climbing$_{(\Omega,\mathcal{N},f_{obj}+f'_{comp})}(s)$
15:             $F'_j \leftarrow f_{obj}(S'_j)$
16:         **end for**
17:         **if** z_score$_{\mathcal{T}}(F',F) \geqslant z_{min}$ **then** // significant score of hypothesis hill_climbing$(f_{obj} + f'_{comp}) \succ_{f_{obj}}$ hill_climbing$(f_{obj} + f_{comp})$
18:             $f_{comp} \leftarrow f'_{comp}$
19:             improve $\leftarrow$ true
20:             nbmut $\leftarrow 0$
21:         **else**
22:             nbmut $\leftarrow$ nbmut $+ 1$
23:         **end if**
24:     **until** improve or (nbmut = maxmut)
25: **until** not improve
26: **return** $f_{alt} := f_{obj} + f_{comp}$

---

4   Experiments

We report in this section preliminary experiments of evolving hill climbers by mutating fitness functions only. Recall that given an objective function f_{obj}, our purpose is to find an alternative fitness function f_{alt} such that a climber guided by f_{alt} will attain, in average, better solutions that a climber guided by f_{obj}. This constitutes a way of determining an alternative fitness landscape, which is more efficiently climbed that the original one.

The following results present experiments on random NK instances of various sizes ($N \in \{128, 256\}$) and ruggedness level ($K \in \{1, 2, 4, 6, 8, 10, 12\}$), which constitute the reference objective functions f_{obj}. First, alternative fitness functions will be generated by summing reference objective functions with evolved complementary functions (see Sect. 3.1). Since the mutation settings depend on the model used, and considering that we expect to favor simplest function models, we choose only NK_N_1 as fitness function model for Algorithm 2 (i.e. NK_128_1 and NK_256_1 complementary functions for tackling NK_128_K and NK_256_K instances respectively). Then we experiment the more general model where the alternative functions are evolved NK_N_1 fitness functions (see Sect. 3.2).

This preliminary set of experiments does not focus on the parameters effects. We choose to fix parameter maxmut to 1000, which appears to be a sufficient value for an efficient fitness function evolution. Selection function for climbers (hill_climbing function in Algorithm 2) is a random first improvement selection heuristic. Parameter p is the sampling size used to evaluate fitness function by means of the statistic test (being the Wilcoxon rank-sum test) and is set to 100.

The initialization function generates a random NK function (from model parameters N and $K = 1$). Algorithm 3 specifies how NK functions are mutated. Mutation parameters M_1 and M_2 are respectively set to 0.05 and 5, which are appropriate values to statistically improve alternative functions with a non-negligible probability.

Algorithm 3. NK fitness function mutation

Input/Output: a NK fitness function f, described with N coefficient tables Γ_i and N ordinal sets of bit interdependencies l_i (of size K).
Parameters: maximal mutation rate ($M_1 \in [0,1]$), maximal number of changed links ($M_2 \in [\![0, K]\!]$).
1: Randomly generate $m_1 \leqslant M_1$ and $m_2 \leqslant M_2$ such that $(m_1, m_2) \neq (0,0)$
2: **for** $m \leftarrow 1$ to $\lceil m_1 \cdot N \cdot (K+1) \rceil$ **do**
3:     Randomly select $i \in [\![1, N]\!]$ and $k \in [\![1, 2^{K+1}]\!]$
4:     Replace Γ_{ik} by a random value in $[0,1]$
5: **end for**
6: **for** $m \leftarrow 1$ to m_2 **do**
7:     Randomly select $i \in [\![1, N]\!]$ and $j \in [\![1, K]\!]$
8:     Replace l_{ij} by a random value in $[\![1, N]\!]$
9: **end for**

For each instance, we ran 100 times Algorithm 2, leading to 100 alternative fitness functions $f_{\text{alt}}^1, \ldots, f_{\text{alt}}^{100}$. Then we perform 100 executions of each climber $HC(f_{\text{alt}}^i)$, to obtain 100 (not necessarily distinct) local optima $s_1^i, \ldots s_{100}^i$ for each function f_{alt}^i. The average quality of the algorithms $HC(f_{\text{alt}}^i)$ is estimated by averaging the objective value of all solutions, i.e. $\sum_i^{100} \sum_j^{100} f_{\text{obj}}(s_j^i)/10000$.

These quality scores are reported in Table 1 (column $HC(f_{obj} + f_{comp})$). We also report in the last column the results obtained when the alternative function is directly mutated (i.e. without using the objective function for hill climbings and therefore with no need of any *complementary* function), with even better results on more rugged instances ($K \leqslant 6$). The main result here is that the average quality of $HC(f_{alt})$ is always better than the average quality of $HC(f_{obj})$, which has been estimated by averaging 10,000 local optima returned by the reference hill climber. $HC(f_{obj} + f_{comp})$ improves further $HC(f_{obj})$ on smoother landscapes, but obtain unexpected poor performances on more rugged instances where f_{comp} seems to penalize the search. This can be explained by the use of inadequate mutation parameters as well as the difficulty to combine two functions (f_{obj} and f_{comp}) of very different ruggedness. However, these preliminary experiments point out the relevance and the complementarity of both approaches. Further parameter analysis and experiments will be dedicated to improving the quality and the adaptability of the evolution strategy.

Table 1. Comparison of hill climbers based on original objective function (f_{obj}) and evolved alternative fitness functions

Instance	$HC(f_{obj})$	$HC(f_{obj} + f_{comp})$	$HC(f_{alt})$
nk-128-1	0.709	**0.723**	0.718
nk-128-2	0.716	**0.745**	0.726
nk-128-4	0.727	**0.765**	0.732
nk-128-6	0.717	0.715	**0.728**
nk-128-8	0.715	0.711	**0.721**
nk-128-10	0.708	0.706	**0.714**
nk-128-12	0.700	0.699	**0.708**
nk-256-1	0.691	**0.705**	0.703
nk-256-2	0.713	**0.735**	0.723
nk-256-4	0.719	0.729	**0.731**
nk-256-6	0.723	0.718	**0.733**
nk-256-8	0.717	0.710	**0.722**
nk-256-10	0.710	0.707	**0.720**
nk-256-12	0.707	0.702	**0.712**

In order to better investigate how an evolved climber explores the original fitness landscape, we compare in Fig. 1 the average f_{obj}-fitness evolution of f_{alt}-guided climbers to f_{obj}-guided ones. The black line represents the average objective value evolution of 10,000 climbers on a reference fitness landscape. The grey line represents the average objective values obtained from climbers executed on 100 alternative fitness landscapes (100 climbers per landscape). For a better understanding, we also indicate, in dotted lines, the evolution of 4 random hill climbings (2 on the reference landscape, 2 on alternative landscapes). Note that a strict hill climbing on alternative landscape naturally provides in general a non-monotonic walk according to an objective function f_{obj}. This illustrates that such climber may cross reference landscape local optima.

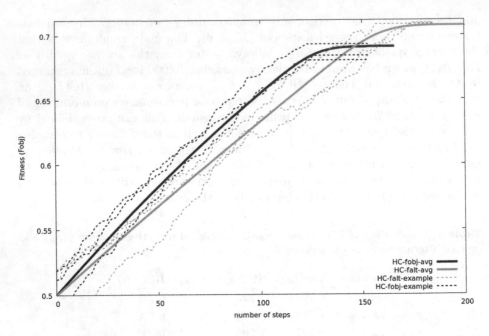

Fig. 1. Performance comparison of alternative and reference fitness functions (hill climbing based evaluation).

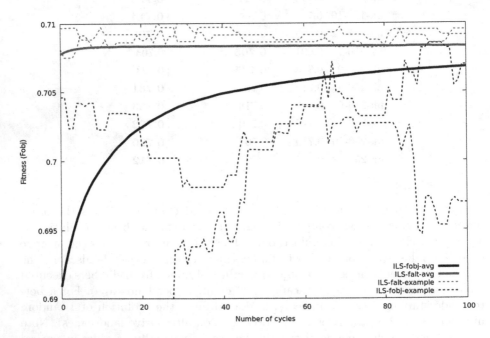

Fig. 2. Comparison of iterated local searches based on alternative and reference fitness functions.

Figure 2 extends the comparison by iterating climbers within an Iterated Local Search scheme. We observe that an iterated search only improves very few the solutions provided by a climber running on an alternative landscape. However, such a climber used alone is more powerful that an iterated local search performed on the original fitness landscape.

5   Conclusion

In this paper, we have proposed a fitness landscape evolution framework whose purpose is to generate better fitness landscapes with regards to a given hill climbing solving technique. Our preliminary experiments show that such an approach may be efficient to improve the ability of a climber to reach better local optima, without involving random perturbations used in more sophisticated local search strategies. Compared to related existing work, our approach is focused on the solving process and does not aim at modifying nor approximating the reference objective function. Moreover, the evolution fitness model may be disconnected from the initial problem model.

This first attempt must of course be improved by considering more different instances with various adjustable fitness properties. The mutation process used in the fitness landscape evolution algorithm must also be finely studied and analyzed.

Our approach can especially be relevant in the context of Black Box optimization problems, where the objective function that must be optimized can be evaluated but some of its properties cannot be effectively used to guide the search. While weighted penalty functions have been successfully used for solving constraint satisfaction or Boolean satisfiability problems [11], they rely on a known model where constraints, variables and their relationships are explicitly defined. As possible application domain, we may point out problems for which the computation of the objective function values is particularly time consuming (e.g., when a simulation or a software is required).

Acknowledgements. This work is partially supported by the *Région Pays de la Loire* through the *Atlantic 2020 programme*.

References

1. Basseur, M., Goëffon, A.: Climbing combinatorial fitness landscapes. Appl. Soft Comput. **30**, 688–704 (2015)
2. Battiti, R., Brunato, M., Mascia, F.: Reactive Search and Intelligent Optimization (2007)
3. Chicano, F., Whitley, D., Alba, E.: A methodology to find the elementary landscape decomposition of combinatorial optimization problems. Evol. Comput. **19**(4), 597–637 (2011)
4. Eiben, A.E., Hinterding, R., Michalewicz, Z.: Parameter control in evolutionary algorithms. Trans. Evol. Comput. **3**(2), 124–141 (1999)

5. Hutter, F., Hoos, H.H., Stützle, T.: Automatic algorithm configuration based on local search. In: AAAI, pp. 1152–1157. AAAI Press (2007)
6. Kauffman, S.A., Weinberger, E.D.: The NK model of rugged fitness landscapes and its application to maturation of the immune response. J. Theoret. Biol. **141**(2), 211–245 (1989)
7. Lobo, F.G., Lima, C.F., Michalewicz, Z.: Parameter Setting in Evolutionary Algorithms, 1st edn. Springer, Heidelberg (2007). https://doi.org/10.1007/978-3-540-69432-8
8. López-ibáñez, M., Dubois-Lacoste, J., Stützle, T., Birattari, M.: The irace package: iterated racing for automatic algorithm configuration. Technical report (2011)
9. Lourenço, H.R., Martin, O.C., Stützle, T.: Iterated local search. In: Glover, F., Kochenberger, G.A. (eds.) Handbook of Metaheuristics. International Series in Operations Research & Management Science, vol. 57, pp. 320–353. Springer, Boston (2003). https://doi.org/10.1007/0-306-48056-5_11
10. Nenad, M., Pierre, H.: Variable neighborhood search. Comput. OR **24**(11), 1097–1100 (1997)
11. Morris, P.: The breakout method for escaping from local minima. In: Proceedings of the Eleventh National Conference on Artificial Intelligence, AAAI 1993, pp. 40–45. AAAI Press (1993)
12. Rice, J.R.: The algorithm selection problem. Adv. Comput. **15**, 65–118 (1976)
13. Vanneschi, L., Pirola, Y., Mauri, G., Tomassini, M., Collard, P., Verel, S.: A study of the neutrality of Boolean function landscapes in genetic programming. Theor. Comput. Sci. **425**, 34–57 (2012)
14. Verel, S., Ochoa, G., Tomassini, M.: Local optima networks of NK landscapes with neutrality. IEEE Trans. Evol. Comput. **15**(6), 783–797 (2011)

Bayesian Immigrant Diploid Genetic Algorithm for Dynamic Environments

Emrullah Gazioglu$^{(\boxtimes)}$ and A. Sima Etaner-Uyar

Istanbul Technical University, Sariyer, Istanbul, Turkey
{egazioglu,etaner}@itu.edu.tr

Abstract. In dynamic environments, the main aim of an optimization algorithm is to track the changes and to adapt the search process. In this paper, we propose an approach called the Bayesian Immigrant Diploid Genetic Algorithm (BIDGA). BIDGA uses implicit memory in the form of diploid chromosomes, combined with the Bayesian Optimization Algorithm (BOA), which is a form of Estimation of Distribution Algorithms (EDAs). Through the use of BOA, BIDGA is able to take into account epistasis in the form of binary relationships between the variables. Experiments show that the proposed approach is efficient and also indicates that exploiting interactions between variables is important to adapt to the newly formed environments.

Keywords: Evolutionary Algorithms · Estimation of Distribution Algorithms · Bayesian Optimization Algorithm · Dynamic environments

1 Introduction

In a dynamic environment, there are a few components in the system that can change over time, i.e. the objectives, constraints and the problem instance itself. Any or all of these components may change. A dynamic environment can be classified as described in [6] by the: (i) Frequency, (ii) Severity, (iii) Predictability and (iv) Cycle length/accuracy of the change that occurs.

One of the main issues in dynamic environments is tracking down the moving optima as closely as possible. Another issue is to catch the change quickly and adapt to the newly formed environment as quickly as possible. There are some strategies proposed for these types of problems [11]: (i) Generating diversity after a change, (ii) Maintaining diversity throughout the run, (iii) Multipopulation approaches [14] (iv) Memory-based approaches [12].

In order to generate or increase diversity after a change, hypermutation [7] and variable local search [22] methods are proposed. These two are techniques that increase the probability of mutation for a while when a change occurs. As an example of an approach that *maintains diversity throughout a run*, Baykasoglu

Supported by organization x.

and Ozsoydan used the *Triggered Random Immigrant (TRIM)* mechanism which introduces a portion of randomly generated new individuals into the population when the diversity level of the population falls under the predefined threshold in [3].

Also, in [1] both hypermutation and random immigrant mechanisms are hybridized to deal with the loss of diversity for dynamic environments.

In spite of having a lot of optimization algorithms in literature for dynamic environments, almost none of them investigates the interactions between variables (genes) in a chromosome. In order to achieve that, the power of both Evolutionary Algorithms (EA) and the Estimation of Distribution Algorithms (EDA) are combined in this study.

This paper continues as follows: The literature summary is mentioned in Sect. 2, then in Sect. 3 the proposed approach BIDGA is explained, in Sect. 4 the experiments are detailed and finally in Sect. 5 results are discussed and the paper is concluded.

2   Literature Summary

2.1   Estimation of Distribution Algorithms

Estimation of Distribution Algorithms (EDAs) consider the optimization problems as a probability theory problem so it learns from the past and acts by using the probability that it has learned. The common feature of all the EDAs is using probabilistic information and generating new solution candidates according to that probabilistic information at each generation of the algorithm.

In [16], the authors classified the EDAs based on the problem decomposition: (i): No dependencies, (ii): Pairwise dependencies, and finally (iii): Multivariate dependencies. We can give the Population-Based Incremental Learning (PBIL) [2] algorithm as an example to the $(i)^{th}$ model. For the *pairwise dependencies*, Mutual Information Maximizing Input Clustering (MIMIC) [9] might be an example. There are number of EDA based approaches, which are proposed in the literature in order to handle dynamism and uncertainty [19,20]. One of them is a dual population PBIL, Uludag et al. [20] proposed a multi-phase hybrid framework, referred to as Hyper-Heuristic based dual population Estimation of Distribution Algorithm (HH-EDA2) that utilizes offline and online learning methods at successive stages for solving dynamic optimization problems.

In this paper, we used the Bayesian Optimization Algorithm with Decision Graphs (dBOA) [15] which falls in the $(iii)^{th}$ model mentioned above. In this model, directed acyclic graphs or undirected graphs are used to show dependencies.

Bayesian Optimization Algorithm

In BOA, at first, the algorithm randomly produces a population and then calculates their fitness values to pick best n of them to create a Bayes Network to exploit which variables are dependent on each other and which are not. Rest of the algorithmic loop, each generation, it re-creates (not updating) the network by using the population from scratch. Since there is no prior knowledge about the variables, it uses a greedy approach to create the network. The most important parameter of BOA is k, which indicates the maximum number of incoming edges to a node in the network. In other words, k determines the maximum number of parents of a node when creating a network. Since it uses a greedy approach at each iteration, it uses the Bayesian Dirichlet equivalence (BDe) metric to measure how good the network is.

In our work, we used BOA for creating a Bayesian Network by using the Diploid Genetic Algorithm's (DGA) population and sampling a sample-population in order to decide the phenotype value of an individual which will be detailed in the proposed method section.

2.2 Memory Based Genetic Algorithms for DOPs

As mentioned before, simple GAs (sGA) quickly converge to a solution but when a change occurs, it loses its genetic diversity. Using memory is one of the solution methods among others for this problem. Memory can be implemented in two ways: *(i) Implicitly* which uses redundant representation, or *(ii) Explicitly* which uses extra memory to save useful information.

In this section, we investigated the GAs proposed for dynamic environments. In the random immigrants GA (RIGA) proposed in [8], a small portion of the individuals are replaced with the randomly generated new individuals to maintain the diversity at each selection step of the algorithm. However, it is observed that RIGA is not a good choice if the severity of change is very low since the individuals from the previous generation may still be useful in the new (current) environment [25]. In order to prevent these kinds of problems, [25] proposed elitism-based GA (EIGA). In EIGA, part of the population is replaced with the mutated copies of the elite solution of the previous generation. Thus, the mutated solutions will have higher chance to catch the new optimum, if the environmental change occurs slightly. However this time, EIGA's performance drops when the severity of change rises [26]. Furthermore, its performance also drops in periodically changing environments. As a solution to this problem, in [27], Yang proposed hybrid GA (HIGA) which combines both RIGA and EIGA. In HIGA, beside the random immigrants, also mutated copies of the elite solution are introduced into the population at each generation.

Besides the above-mentioned kind of techniques, there are also memory-based GAs have been proposed in recent years. In [25], Yang proposed memory/search GA (MSGA) as a peer GA proposed by Branke in [4,5]. In MSGA there are two populations: a search population and a memory population. First, the sizes of the populations are equal. At each generation, their size is updated according to their performance: Each generation, the better population gets more space from the total population size. The memory population saves the solutions and the search population is initialized again if a change is detected. Besides the MSGA, Yang also proposed memory-enhanced GA (MEGA) in [26]. In MEGA, if an environmental change is detected, both populations are merged and the best $n\%$ individuals of the merged population go into the genetic operators to construct a new population and memory remains the same. In [26], another hybrid GA, namely memory and random immigrant GA (MRIGA) is proposed. The only difference of MRIGA from MEGA is that MRIGA replaces the worst solutions with the random immigrants. Apart from them, in [24] memory and elitism-based immigrants GA (MIGA) is proposed. MIGA uses the same memory updating mechanism just like in the MSGA, MEGA, and MRIGA. Different from the three of them, in MIGA, at each generation memory is reevaluated and the best ones are retrieved to produce immigrants by using the bitwise mutation with a certain probability. Next, these new individuals replace the worst solutions in the population. In [18], Qian et al. proposed the Environment Reaction GA (ERGA). ERGA is an explicit memory using algorithm like MIGA, but it has two different features from MIGA: *(1)* While MIGA starts with a randomly initialized memory, ERGA starts with an empty memory. The memory updating method in ERGA works as follows. First, the elite individual in the current generation is saved to the memory if there isn't any other same solution as the elite one in the memory. Secondly, if the memory's size is due to the predefined size ($memSize$), the elite one replaces the most similar solution in the memory. However, in MIGA, the elite solution replaces any randomly initialized solution in the memory if there still exists any. *(2)* In ERGA, both random immigrants and mutated solutions in the memory are used to handle the new environment. On the other hand in MIGA, this action is performed in every generation.

An example to the implicit memory methods, in [21], the proposed diploid algorithm domGA uses a probability vector (probability of being 1) to decide the value of a phenotype's variable if its genotypes' variables are different. Note that this approach works without considering interactions (called epistasis) between variables.

3   Proposed Approach

In order to efficiently solve the DOPs and adapt to the newly formed environment as fast as possible, in this paper, we propose Bayesian Immigrant GA (BIDGA) which is formed via injecting dBOA into DGA. In the algorithm, while all the optimization process is carried out with the DGA, the dBOA is used for the phenotype decision mechanism. In order to embed the dBOA in our work, the source code of the dBOA is downloaded from the author's web page[1] and used with a small interface implementation. Also, a small portion of the population sampled via dBOA is used as an immigrant population. The general pseudo-code of the BIDGA can be seen in Algorithm 1.

Besides BIDGA, to show the effect of the diploidy in the plots, we also implemented the haploid version of BIDGA which is called Bayesian Immigrant GA (BIGA). In BIGA, of course, there is no phenotype decision mechanism, instead, at each generation, a portion of the randomly selected population is replaced with the samples from the sample-population of dBOA. The rest of the

---

[1] http://martinpelikan.net/software.html.

algorithm is the same as the simple GA in BIGA: tournament selection, uniform crossover, bitwise mutation, and elitism-1.

3.1   Explanation of BIDGA

In the literature summary section, it can be seen that different types of GAs for DOPs were investigated in [18,21,26]. However, none of them considers the interactions between variables in a solution candidate (chromosome) except PBIL-based approaches which are not GA. As we know, in the real world, things affect each-other and almost nothing is independent of nature. For this reason, we injected dBOA into the DGA to exploit the dependencies between the variables and take action according to this outcome.

First of all, in BIDGA, a diploid representation is used, which means that each individual in the population has two genotypes and one phenotype. The aim of using diploidy is to have an implicit memory scheme to carry the knowledge learned so far to the next generations. In order to make that happen, all the genetic operators are only applied to the genotypes of the individuals. Phenotype is used for only fitness calculations. Thus, the fitness values of the individuals don't affect the genotypes of the individuals directly.

At the beginning of the algorithm, the genotypes of the individuals in the population are created randomly and phenotype values are determined randomly if corresponding genotype values are different (if equal, use that value). After that, fitness values are calculated and then a Bayes Network (BN) is constructed using the best $k\%$ individuals of the population. Then, the BN is used for sampling a sample-population with the same size of BIDGA's population. Finally, a probability vector is constructed by using this sample-population just like PBIL does. This vector is used as the decision method for individuals' phenotype values if the corresponding genotype values are different.

In the selection step of the BIDGA, the typical tournament selection is used. Besides that, a randomly selected portion of the sample-population is immigrated to the main population to achieve immigration. Here, the immigrated samples are copied to both genotypes of the individuals. Also, the best individual from the previous generation is copied to the current population (elitism) without re-calculating its fitness value.

For crossover operation, the uniform crossover method is used. In the crossover function, a binary crossover mask vector is randomly created and then at each iteration, two different individuals are randomly selected, say i_1, i_2. Then with a probability of p_c, for each variable (gene), j, the first one's (i_1) first genotype's j^{th} variable is copied to the second one's (i_2) second genotype's j^{th} variable and second one's second genotype's j^{th} variable is copied to the first one's first genotype's j^{th} variable if the randomly generated binary crossover mask vector's corresponding value is 1.

For mutation, the well-known bitwise mutation operator is used. In the bitwise mutation operation, the genotype values of each individual in the population are flipped with a probability of p_m.

After the genetic operators, phenotype construction is applied. As shown in Algorithm 1, if the genotype values are the same, that value is used for phenotype also. Otherwise, dBOA's probability vector is used to decide what it is going to be; zero or one.

Algorithm 1: Pseudocode of BIDGA

$population \leftarrow initialize(populationSize)$;
$evaluate(population)$;
$BN \leftarrow constructBayesNetwork(population)$;
$samplePopulation \leftarrow sampleBayesNetwork(BN)$;
$BOAprobVector \leftarrow constructProbVector(samplePopulation)$;
while *termination condition not met* **do**
  $tournamentSelection(population)$;
  $uniformCrossover(population)$;
  $bitwiseMutation(population)$;
  $genotype2phenotype(population)\{$
    if $genotype_j^1 = genotype_j^2$ *then* $phenotype_j \leftarrow genotype_j^1$;
    else;
      $p \leftarrow rand()$;
      $phenotype_j \leftarrow (p < BOAprobVector_j)? : 1 : 0$;
  $\}$
  $evaluate(population)$;
  if $modulo(generation, \tau) = 0$ *then* $changeEnvironment()$;
  $constructBayesNetwork(population)$;
  $samplePopulation \leftarrow sampleBayesNetwork(BN)$;
  $BOAprobVector \leftarrow constructProbVector(samplePopulation)$;

---

4   Experimental Design

4.1   Creating Dynamic Environments

The XOR Generator proposed in [23] is a dynamic environment generator with different difficulty degrees for any binary encoded stationary problem. Assume $\overrightarrow{X}$ is a binary encoded solution candidate for a problem, then the fitness value of that solution candidate is calculated for time t as shown in Eq. 1.

$$f(\overrightarrow{X}, t) = f(\overrightarrow{X} \oplus \overrightarrow{M}_k) \qquad (1)$$

where $\oplus$ is the XOR operator and $\overrightarrow{M}_k$ is a masking vector for k^{th} environment. At the beginning, mask $\overrightarrow{M}$ is initialized with zero. After that, every τ generations, it is updated as shown in Eq. 2.

$$\overrightarrow{M}_k = \overrightarrow{M}_{k-1} \oplus \overrightarrow{T}_k \qquad (2)$$

where $\overrightarrow{T}_k$ is a binary template.

Besides the above generator (*random environment*), there are also *cyclic* and *cyclic with noise* environments which are proposed in [28]. In order to construct a cyclic environment, we first construct K binary encoded templates[2] $\overrightarrow{T}(0), ..., \overrightarrow{T}(K-1)$ randomly but exclusively: Assume that each template is a row of a matrix, then the number of total ones in a column of the matrix must be 1. Assume $\overrightarrow{M}(0)$ is composed of zeros, then the rest of the XORing masks are constructed as shown in Eq. 3.

$$\overrightarrow{M}(i+1) = \overrightarrow{M}(i) \oplus \overrightarrow{T}(i\%K), \qquad i = 0, ...2K-1. \tag{3}$$

With the above formula, after K environmental change the $\overrightarrow{M}(K)$ will be all ones and then the K base states will be reused to construct next K masks till to return to the environment $\overrightarrow{M}(0) = \overrightarrow{0}$.

By using the above cyclic environment generator, Yang and Yao also introduced the cyclic environment with noise [28] via introducing noise to the next mask by using bitwise flipping with a small probability, called p_n.

4.2 Problems for Testing BIDGA

Decomposable Unitation-Based Functions
Decomposable Unitation-Based Functions (DUFs) [17] have been widely used in DOPs since they have different difficulty levels and have different dependency degrees. All of the functions are composed of 4-bit building blocks. Each block's fitness value, $u(x)$, is calculated separately and at the end, all of the $u(x)$ values are added to each other to obtain the general fitness value of the individual as follows: $f(x) = \sum_{i=1}^{l/4} u(x)$, where l is the length of a chromosome.

DUF1 is the simple One-Max problem that aims to maximize the number of 1s in a chromosome. In DUF2, also known as the Plateau function, the optimum value is surrounded by the sub-optimum values. DUF3 (a.k.a. Deceptive function) is a very hard problem because of having low-order building blocks. DUF4 (a.k.a. Royal Road function) requires a full of ones in the building block to return a fitness value different than zero [28].

In this study, dynamic DUFs (DDUF) are constructed by using four stationary DUFs (DUF1, DUF2, DUF3, DUF4). By using each DUF, three dynamic DUFs are constructed as suggested in [28] which are cyclic, cyclic with noise, and random.

Dynamic Knapsack Problem
Dynamic Knapsack Problem (DKP) [13] is the dynamic version of the well-known 0-1 Knapsack Problem which aims to collect a number of items to increase the profit while the total weight of the collected items doesn't exceed the given capacity C. The formulation of 0-1 Knapsack Problem is given in Eq. 4:

---

[2] Each template should contain $\rho \times l = l/K$ ones.

$$max \quad f(x) = \sum_{i=1}^{n} p_i * x_i \tag{4}$$

$$subject\ to \sum_{i=1}^{n} p_i x_i \leq C \quad x_i \in \{0,1\},$$

where x_i is the binary decision variable whether item i is taken or not, p_i is the profit of item i, w_i is the weight of item i and finally, C is the total capacity.

As it is indicated in [13], because the difficulty of the DKP is affected by the correlation between weights and profits, we generated highly correlated datasets for our study via the following way: First, create a weight vector by $w_i = rand(1, v)$, then a profit vector by $p_i = w_i + r$, and determine the capacity with $C = 0.5 \times \sum_{i=1}^{n} w_i$, where, n is the number of items, $rand(1, v)$ is a uniformly random value between 1 and v. The parameters v and r are set to 100 and 50 respectively as previously done in [18].

4.3 Dynamic Environments

In order to test the proposed method BIDGA, mainly two groups, but in total seven different problems are implemented. Four of them are DUF problems from 1 to 4 and three of them are Dynamic Knapsack Problems (DKP) that are proposed in [13] with 100, 250 and 500 items respectively.

For testing BIDGA on DDUFs, the very same test environment prepared by Yang and Yao [26] is used. For testing and comparing the DKP results, parameters are set to the same experimental design in [18]: The number of generation, $G = 200 \times \tau$, item sizes are 100, 250 and 500, frequency (τ) is set to 20 and 40 respectively.

For each problem, the environment is changed at every τ generations where τ is set to 10 and 25 for DDUFs and 20 and 40 for DKPs. The severity of the change, ρ, is set to 0.1, 0.2, 0.5, 1.0 for all the problems. By using these severity values in the previously mentioned equation, $\rho \times l = l/K$, there will be four different base states which are 2, 4, 10 and 20. For the cyclic environment with noise, the noise probability p_n is set to 0.05.

In total, 24 Dynamic DUF (DDUF) problems are constructed using 4 values of ρ, 2 values of τ and 3 different environments for each DDUF and each DDUF is solved by BIDGA, BIGA, domGA, and sGA separately.

For each test case, 50 independent runs were executed with the same set of seeds. For each run, the best-of-generation is saved and overall performance [28] is measured as shown in Eq. 5.

$$\overline{F}_{BOG} = \frac{1}{G} \sum_{i=1}^{G} \left(\frac{1}{N} \sum_{j=1}^{N} F_{BOG_{ij}} \right) \tag{5}$$

where G is the number of generation for each run, N is the number of runs which is 50, $F_{BOG_{ij}}$ is the best of generation fitness value of generation i, of run j and $\overline{F}_{BOG}$ is the overall offline performance.

Table 1. Parameter settings of BIDGA and BIGA

Parameter	On DDUFs	On DKPs
G	5000	$\tau \times 200$
Population size	100	100
Chromosome length	100	**100**, 250, 500
p_c	1.0	1.0
p_m	0.05, **0.08**, 0.1, 0.15	0.05, **0.08**, 0.1
β	**0.1**, 1.0	0.1
p_n	0.05	0.05
Tournament size	4	4

Preliminary Tests for Parameter Tuning

In order to see the effect of the Bayesian immigrant mechanism of BIDGA, its introducing rate β is set to 0.1 and 1.0 respectively. Finally, to see the effect of the mutation, the p_m is set to 0.05, 0.08, 0.1 and 0.15 for DDUFs and it is set to 0.05, 0.08 and 0.1 for DKPs respectively. All the parameter settings can be seen in Table 1. Note that in the table, only the bold ones are shown in the plots in this study because of the space limitation.

BIDGA is tuned under the different parameter settings (for β and p_m) as mentioned previously in the Dynamic Environment section. In [10], we compared the dBOA's β rate for 0.1 and 1.0. Results show that using a high rate of β causes to stuck in local optima since the individuals in the sample-population of dBOA are highly similar to each other. For this reason, β is set 0.1 for further processes. Again, in [10], by fixing β is to 0.1, the mutation rate is investigated next. While on DDUFs tuning mutation rate doesn't show a significance, on DKPs it is clearly seen that increasing the mutation rate increases the performance. In this case, while the 0.05 mutation rate performs the worst, 0.08 is slightly better than 0.1. For this reason, p_m is set to 0.08 for further processes.

5   Results and Conclusion

In this section, we are going to compare and contrast BIDGA, BIGA, domGA, and sGA on DDUFs and compare BIDGA's performance on DKP-100 dataset to the results given[3] in [18]. Due to the space limitation, only the results with $\tau = 10$ will be given for DDUFs and $\tau = 20$ will be given for DKP-100 on the plots. The other results (different τ values, DKP-250, DKP-500) can be seen in [10].

There are several conclusions that can be obtained from the results of the experiments on the DDUFs. Firstly, as seen in Fig. 1, the Bayesian immigrant algorithms (BIDGA and BIGA) perform better than the rest on most of the severity levels. This shows the importance of exploiting interactions between the variables. Also, it can be observed from BIDGA and BIGA that, using diploidy

---

[3] The exact results given in Table 2 in [18].

(implicit memory) increases the performance of the algorithm. Except DDUF1 (simple One-Max), the rest of the problems have interactions between the variables as mentioned before. For this reason, in the results, it can be observed that although all the algorithms show almost the same trend, on the other DDUFs, BIDGA is better than the other algorithms and doesn't show the same trend with them.

In order to see the effect of the Bayesian decision mechanism, consider the BIDGA and domGA in the results. Almost on each case, BIDGA adapts to the newly formed environment better than the domGA. This is because while dBOA exploits interactions between the variables, domGA's probability vector doesn't. Also to see the effect of the diploidy mechanism, consider the BIDGA and BIGA. Again, at each case, BIDGA performs better than the BIGA. This tells us the effect of using implicit memory.

Viewing the results from left to right in Fig. 1, it can be seen that performances of the algorithms relatively decreases since the noise effect reduces the precision of cycles. Viewing the results from the top to bottom in Fig. 1, the performance of DDUF1 vs. rest, it can be seen that differences of performances are increase between algorithms because of the epistatic problem structure of DDUF2, DDUF3, and DDUF4. As the interactions of the variables increase, the performances of the algorithms except BIDGA are decreased.

The statistical results of the comparing BIDGA vs. others by two-tailed t-test at a 0.05 level of significance are given in Table 2. In the table "$s+$" means BIDGA is significantly better than the other algorithm, "$+$" means BIDGA is better than the other, "$s-$" means BIDGA is significantly worse than the other algorithm and finally "$-$" means BIDGA is worse than the other algorithm as defined in [25].

Besides the above results, when the BIDGA results in Fig. 1 compared to the MIGA results given in [25], except the cyclic environment, BIDGA performs better than the MIGA. Also, in most cases, when the severity is 1.0, MIGA performs better than the BIDGA. This is because MIGA uses an explicit memory to save and reuse the good solutions of the already visited environments.

In Fig. 2, the experimental results on DKP of BIDGA and the results of the other algorithms which are given in [18] can be seen. When we look at the results, it is clearly seen that BIDGA performs better than any other algorithm. In order to understand that, we calculated the averaged Infeasible individual rate in the population just before the environmental change and just after the environmental change and plotted the results on Fig. 3 for $\tau = 20$ case.

From Fig. 3 it is seen that the diversity of the population is well enough to handle the environmental changes. However, for the $\rho = 1.0$ we see that the performance of BIDGA decreases on the Cyclic and Random environment. This is because the nature of the knapsack problem, because dropping good items and taking poor items is highly influential for a knapsack problem. On the other hand, on the Cyclic with noise environment, differences of infeasible rate between *before* and *after* are not too high since the noise effect prevents the population to oscillate between two distinct environments and gives chance to diversification.

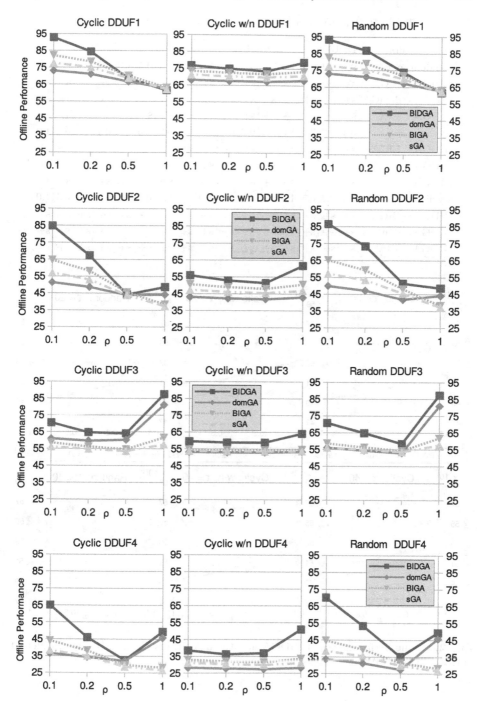

Fig. 1. Offline performances of BIDGA, BIGA domGA and sGA with different options on DDUF1, DDUF2, DDUF3 and DDUF4 under cyclic, cyclic with noise and random dynamic environments with $\beta = 0.1$, $p_m = 0.08$, $\tau = 10$ at $\rho = 0.1$, 0.2, 0.5 and 1.0

Table 2. The t-test results of the BIDGA vs. others at $\tau = 10$

t-test results	DDUF1				DDUF2				DDUF3				DDUF4			
Random, $\rho \Rightarrow$	0.1	0.2	0.5	1.0	0.1	0.2	0.5	1.0	0.1	0.2	0.5	1.0	0.1	0.2	0.5	1.0
BIDGA – BIGA	s+	s+	s+	s−	s+	s+	s+	s+	s+	s+	s+	s+	s+	s+	s+	s+
BIDGA – domGA	s+	s+	s+	s−	s+	s+	s+	s+	s+	s+	s+	s+	s+	s+	s+	s+
BIDGA – sGA	s+	s+	s+	s−	s+	s+	s+	s+	s+	s+	s+	s+	s+	s+	s+	s+
Cyclic, $\rho \Rightarrow$	0.1	0.2	0.5	1.0	0.1	0.2	0.5	1.0	0.1	0.2	0.5	1.0	0.1	0.2	0.5	1.0
BIDGA – BIGA	s+	s+	s−	s−	s+	s+	s−	s+	s+	s+	s+	s+	s+	s+	s+	s+
BIDGA – domGA	s+	s+	s+	s−	s+	s+	s+	s+	s+	s+	s+	s+	s+	s+	+	s+
BIDGA – sGA	s+	s+	s−	s−	s+	s+	s+	s+	s+	s+	s+	s+	s+	s+	s+	s+
Cyclic w/n, $\rho \Rightarrow$	0.1	0.2	0.5	1.0	0.1	0.2	0.5	1.0	0.1	0.2	0.5	1.0	0.1	0.2	0.5	1.0
BIDGA – BIGA	s+	s+	s+	s+	s+	s+	s+	s+	s+	s+	s+	s+	s+	s+	s+	s+
BIDGA – domGA	s+	s+	s+	s+	s+	s+	s+	s+	s+	s+	s+	s+	s+	s+	s+	s+
BIDGA – sGA	s+	s+	s+	s+	s+	s+	s+	s+	s+	s+	s+	s+	s+	s+	s+	s+

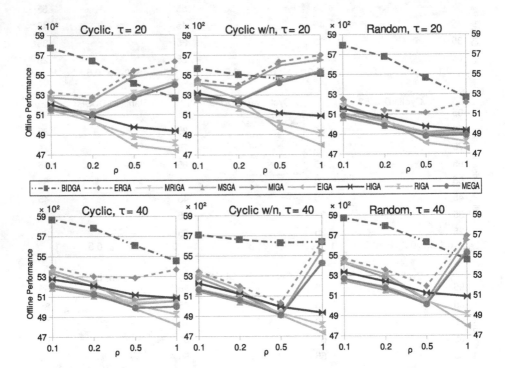

Fig. 2. Offline performances of BIDGA and the algorithms given in [18] with different options on DKP-100

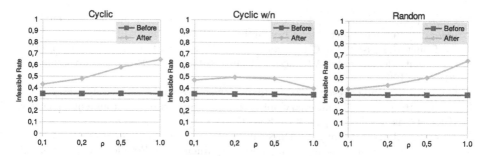

Fig. 3. Infeasible rates of the BIDGA's population at each change (before and after) for $\tau = 20$ on DKP-100

5.1 Conclusion and Future Work

In this study, a new EDA-included GA (BIDGA) is proposed to handle dynamic environments in optimization problems. In the algorithm, the power of the probability theory and a nature-inspired algorithm (GA) is combined and the GA part is implemented with diploid representation to have an implicit memory. The use of a BN to calculate the phenotypes allows interactions between the genes to be taken into account. This feature is seen to be useful especially in the cases where there is epistasis. Experiments are conducted on three types of change dynamics, namely cyclic, cyclic with noise and random and results are compared with similar algorithms taken from literature. The results are very promising to investigate this new type of algorithm further.

References

1. Akandwanaho, S.M., Viriri, S.: A spy search mechanism for memetic algorithm in dynamic environments. Appl. Soft Comput. **75**, 203–214 (2019)
2. Baluja, S.: Population-based incremental learning. A method for integrating genetic search based function optimization and competitive learning. Technical report, Carnegie-Mellon Univ. Pittsburgh, PA, Dept. of Computer Science (1994)
3. Baykasoğlu, A., Ozsoydan, F.B.: Dynamic optimization in binary search spaces via weighted superposition attraction algorithm. Expert Syst. Appl. **96**, 157–174 (2018)
4. Branke, J.: Memory enhanced evolutionary algorithms for changing optimization problems. In: Proceedings of the 1999 Congress on Evolutionary Computation-CEC99 (Cat. No. 99TH8406), vol. 3, pp. 1875–1882. IEEE (1999)
5. Branke, J.: Optimization in dynamic environments. In: Branke, J. (ed.) Evolutionary Optimization in Dynamic Environments, vol. 3, pp. 13–29. Springer, Boston (2002). https://doi.org/10.1007/978-1-4615-0911-0_2
6. Branke, J.: Evolutionary Optimization in Dynamic Environments, vol. 3. Springer, Boston (2002). https://doi.org/10.1007/978-1-4615-0911-0
7. Cobb, H.G.: An investigation into the use of hypermutation as an adaptive operator in genetic algorithms having continuous, time-dependent nonstationary environments. Technical report, Naval Research Lab Washington DC (1990)

8. Cobb, H.G., Grefenstette, J.J.: Genetic algorithms for tracking changing environments. Technical report, Naval Research Lab Washington DC (1993)
9. De Bonet, J.S., Isbell Jr, C.L., Viola, P.A.: MIMIC: finding optima by estimating probability densities. In: Advances in Neural Information Processing Systems, pp. 424–430 (1997)
10. Gazioglu, E.: Bidga results (2019). https://web.itu.edu.tr/egazioglu/bidga/
11. Jin, Y., Branke, J.: Evolutionary optimization in uncertain environments-a survey. IEEE Trans. Evol. Comput. **9**(3), 303–317 (2005)
12. Mavrovouniotis, M., Yang, S.: Direct memory schemes for population-based incremental learning in cyclically changing environments. In: Squillero, G., Burelli, P. (eds.) EvoApplications 2016. LNCS, vol. 9598, pp. 233–247. Springer, Cham (2016). https://doi.org/10.1007/978-3-319-31153-1_16
13. Michalewicz, Z., Arabas, J.: Genetic algorithms for the 0/1 knapsack problem. In: Raś, Z.W., Zemankova, M. (eds.) ISMIS 1994. LNCS, vol. 869, pp. 134–143. Springer, Heidelberg (1994). https://doi.org/10.1007/3-540-58495-1_14
14. Ozsoydan, F.B., Baykasoğlu, A.: Quantum firefly swarms for multimodal dynamic optimization problems. Expert Syst. Appl. **115**, 189–199 (2019)
15. Pelikan, M.: The Bayesian optimization algorithm (BOA) with decision graphs. IlliGAL Report (2000025) (2000)
16. Pelikan, M., Hauschild, M.W., Lobo, F.G.: Estimation of distribution algorithms. In: Kacprzyk, J., Pedrycz, W. (eds.) Springer Handbook of Computational Intelligence, pp. 899–928. Springer, Heidelberg (2015). https://doi.org/10.1007/978-3-662-43505-2_45
17. Peng, X., Gao, X., Yang, S.: Environment identification-based memory scheme for estimation of distribution algorithms in dynamic environments. Soft. Comput. **15**(2), 311–326 (2011). https://doi.org/10.1007/s00500-010-0547-5
18. Qian, S., Liu, Y., Ye, Y., Xu, G.: An enhanced genetic algorithm for constrained knapsack problems in dynamic environments. Natural Comput. **18**(4), 913–932 (2019). https://doi.org/10.1007/s11047-018-09725-3
19. Uludağ, G., Kiraz, B., Etaner-Uyar, A.Ş., Özcan, E.: A framework to hybridize PBIL and a hyper-heuristic for dynamic environments. In: Coello, C.A.C., Cutello, V., Deb, K., Forrest, S., Nicosia, G., Pavone, M. (eds.) PPSN 2012. LNCS, vol. 7492, pp. 358–367. Springer, Heidelberg (2012). https://doi.org/10.1007/978-3-642-32964-7_36
20. Uludağ, G., Kiraz, B., Etaner-Uyar, A.Ş., Özcan, E.: A hybrid multi-population framework for dynamic environments combining online and offline learning. Soft. Comput. **17**(12), 2327–2348 (2013). https://doi.org/10.1007/s00500-013-1094-7
21. Uyar, A.Ş., Harmanci, A.E.: A new population based adaptive domination change mechanism for diploid genetic algorithms in dynamic environments. Soft. Comput. **9**(11), 803–814 (2005). https://doi.org/10.1007/s00500-004-0421-4
22. Vavak, F.: Adaptive combustion balancing in multiple burner boiler using a genetic algorithm with variable range of local search. In: 7th International Conference on Genetic Algorithm. Morgan Kaufmann (1997)
23. Yang, S.: Constructing dynamic test environments for genetic algorithms based on problem difficulty. In: 2004 Congress on Evolutionary Computation (CEC 2004), vol. 2, pp. 1262–1269. IEEE (2004)
24. Yang, S.: Memory-based immigrants for genetic algorithms in dynamic environments. In: Proceedings of the 7th Annual Conference on Genetic and Evolutionary Computation, pp. 1115–1122. ACM (2005)

25. Yang, S.: Genetic algorithms with elitism-based immigrants for changing optimization problems. In: Giacobini, M. (ed.) EvoWorkshops 2007. LNCS, vol. 4448, pp. 627–636. Springer, Heidelberg (2007). https://doi.org/10.1007/978-3-540-71805-5_69

26. Yang, S.: Genetic algorithms with memory-and elitism-based immigrants in dynamic environments. Evol. Comput. **16**(3), 385–416 (2008)

27. Yang, S., Tinós, R.: A hybrid immigrants scheme for genetic algorithms in dynamic environments. Int. J. Autom. Comput. **4**(3), 243–254 (2007). https://doi.org/10.1007/s11633-007-0243-9

28. Yang, S., Yao, X.: Population-based incremental learning with associative memory for dynamic environments. IEEE Trans. Evol. Comput. **12**(5), 542–561 (2008)

Ant Colony Optimization Algorithm for a Transportation Problem in Home Health Care with the Consideration of Carbon Emissions

Hongyuan Luo, Mahjoub Dridi, and Olivier Grunder[✉]

Nanomedicine Lab, Univ. Bourgogne Franche-Comté, UTBM, 90010 Belfort, France
{hongyuan.luo,mahjoub.dridi,olivier.grunder}@utbm.fr

Abstract. Home health care (HHC) companies provide the care service for the patients at their homes in order to help them recover from illness or injury. Since transportation cost is one of the largest operating costs in the daily activities of HHC company, it is crucial to optimize daily traveling routes of the HHC vehicles in order to reduce the transportation cost meanwhile improving the service quality to patients. However, transportation has serious impacts on the environment. Therefore, it compels managers of the HHC companies to pay more attention to CO_2 emissions when designing the daily logistics activities. This study addresses a daily transportation problem of a HHC company with the constraints of synchronized visits and carbon emissions. In order to solve the studied problem, we develop an ant colony optimization (ACO) algorithm. The experimental results highlight the efficiency of the proposed ACO algorithm compared with the Gurobi solver with a time limit of 3600 s.

Keywords: Ant colony optimization · Home health care · Synchronized visits · Carbon emissions

1 Introduction

Home health care (HHC) company provides the health care service for the patients at their homes in order to help them recover from illness or injury. According to a survey of the HHC companies, the HHC company conducts various logistic activities including delivering the caregivers, drugs, medical devices from the HHC company (i.e. the depot) to the patients, and biological samples (such as blood and urine) from the patients' homes to the medical laboratory for testing every day [1]. The daily scheduling of the caregivers has been demonstrated to be a very difficult problem but a crucial decision activity for a HHC company [2]. A large number of patients who need care service are usually distributed in a town, a village or a city. Each patient has a different service time horizon (also called time window in the paper) and a different service requirement. Based on the requirement of the patient, the care service may be accomplished by more than one caregiver at the same time. In addition, we assume

© Springer Nature Switzerland AG 2020
L. Idoumghar et al. (Eds.): EA 2019, LNCS 12052, pp. 136–147, 2020.
https://doi.org/10.1007/978-3-030-45715-0_11

that some drugs have their own volume, thus the vehicle capacity is also taken into consideration in this paper. Therefore, in this case, the HHC scheduling problem is similar to a vehicle routing problem with time window (VRPTW) and synchronized visits constraints [3].

As for a HHC company, transportation cost is one of the largest operating costs in company daily activities, thus it is crucial to optimize daily traveling routes of the HHC vehicles in order to reduce the transportation cost meanwhile improving the service quality to patients. However, transportation has serious impacts on the environment, such as resource consumption, toxic effects on ecosystems and humans, noise, and the effect induced by greenhouse gas (GHG) emissions. Among these, GHG, especially carbon dioxide (CO_2) emissions, are the most concerning because CO_2 emissions have direct influences on people's health [4]. If logistics is not scheduled well, it will cause congestion and a large amount of CO_2 emissions. Therefore, it compels managers of the HHC companies to pay more attention to CO_2 emissions when designing the daily logistics activities.

Recently, some scholars have started studying the HHC problems with the consideration of the carbon emissions. Fathollahi-Fard et al. [5] studied the problem of the delivery the required drugs and other HHC services to patients. They firstly introduced the environmental pollution or green emissions into the HHC problems, and developed a bi-objective optimization model. Four fast heuristics are proposed to solve the problem. Xiao et al. [6] also considered the carbon emissions in the HHC transportation problem. They used a capacity VRP (CVRP) model to describe the HHC scheduling problem and proposed an improved cuckoo search (ICS) algorithm for the problem.

This study addresses a daily routing and scheduling problem of a HHC company with the constraints of synchronized visits and carbon emissions. In order to solve the large scale instances, an ant colony optimization (ACO) algorithm is developed. The rest of this paper is organized as follows. Section 2 introduces the scheduling problem and Sect. 3 builds the mathematical model. Section 4 develops an ACO algorithm in order to solve the problem. The computational experiments are described in Sect. 5. Section 6 concludes the paper.

2   Problem Description

This paper addresses a daily routing and scheduling problem of a HHC company with the constraints of synchronized visits and carbon emissions. The problem can be defined as follows. Let $G = (N, A)$ be a directed graph with a set of nodes $N = \{0, 1, ..., n, n+1\}$ and a set of arcs $A = \{(i,j) \, | \, i,j \in N, i \neq j\}$. Node 0 and node $n+1$ represent the depot and the medical laboratory, respectively. Nodes $P = \{1, 2, ..., n\}$ represent the patients who need care service from the HHC company.

Each patient $i \in P$ has a drug and service demand q_i, and each caregiver has the same load and service capacity Q. Each patient $i \in P$ is associated with a service duration τ_i. Each patient $i \in P$ has a service time window $[a_i, b_i]$, where

a_i represents the earliest time and b_i represents the latest time for visiting the patients. Each caregiver is allowed to arrive before the earliest time a_i, but the caregiver must wait until that the time is available for the patient. The caregiver is prohibited to arrive after the latest time b_i. The depot and the laboratory have the same time window, meaning the caregivers must leave from the depot and return to the laboratory between the earliest time and latest time.

Some patients may need synchronized services, which means that two or more caregivers must service these patients simultaneously. In this paper, we only consider two caregivers visit a patient simultaneously. For each patient $i \in P$ with synchronized services, a fictive patient i' who has the same locations, demand, service duration and time window with patient i is generated. We refer all fictive patients to P_f. Therefore, we can define $N' \leftarrow N \cup P_f$, $P' \leftarrow P \cup P_f$, $A' = \{(i,j) \,|\, i,j \in N', i \neq j\}$. We adopt $(i,j) \in P^{sync}$ to represent a couple of patients $i,j \in P'$ who need synchronized services. In other words, i and j are associated to the same patient and must be serviced by two different caregivers simultaneously.

The distance between patient i and j is denoted as d_{ij}. This paper considers the constraints of the carbon emission. Speed has a great influence on carbon emission. Therefore, the speed parameter is employed in the paper. The speed of the vehicle k associated to the caregiver k is v. Based on the speed v, it is very easy to calculate the travel time between i and j. The travel time between i and j is d_{ij}/v.

The problem is developed to determine a set of routes in order to minimize the carbon emissions under the constraints of time windows, capacity and synchronized visits, and the following assumptions: (1) each caregiver has the same service capacity and is associated to a vehicle; (2) each vehicle leaves from the depot and returns to the laboratory, and visits each node at most once; (3) the unused vehicles are assumed to start from the depot and end at the laboratory, in order to prevent from adding the emission cost, we assume that the distance from depot to laboratory is 0; (4) because there are many uncertain factors in the city transportation, the speed of the vehicle is assumed to be a constant average speed; (5) for the patient with synchronized visit services requirement, a fictive patient who has the same locations, demand, service duration and time windows is generated. We assume that the patient at most needs two caregivers to service at the same time; (6) for the patient with synchronized visit services requirement, if caregiver 1 arrives earlier than caregiver 2, caregiver 1 must wait for caregiver 2 and then serving the patient together.

3   Mathematical Formulation

In this section, we will introduce the mathematical model. Firstly, we introduce the theory of carbon emissions; then, a mixed-integer programming (MIP) model is developed for this problem.

3.1   Carbon Emissions

This paper adopts the emissions function developed by the United Kingdom Transport Research Laboratory. The emissions function has been used by many researchers, such as [7–9], and so on, which can demonstrate the effectiveness of the emission function. The emissions function $\varepsilon\,(v)$ is provided as follows:

$$\varepsilon\,(v) = L + av + bv^2 + cv^3 + dv^{-1} + ev^{-2} + fv^{-3} \qquad (1)$$

where v is the speed of the vehicle in kilometer(km)/hour(h), and the coefficients L, a, b, c, d, e and f will be different under the vehicles with different types and sizes. The coefficients are adopted the settings in [9], and the values of L, a, b, c, d, e and f are 765, -7.04, 0, 0.006320, 8334, 0, 0, respectively.

The vehicle will emit $\varepsilon\,(v)$ gram(g)/km carbon dioxide (CO_2) when the vehicle is driven at the speed v. Therefore, the CO_2 emission of a vehicle travels from patient i to patient j can be expressed as:

$$E_{ij} = \varepsilon\,(v)\,d_{ij} \qquad (2)$$

where the units of E_{ij} and d_{ij} are g and km, respectively.

As is shown in Eq. (1), it is very clear that the CO_2 emissions rate $\varepsilon\,(v)$ will vary with different speed. Therefore, an optimal speed can be found in order to reduce the CO_2 emissions. However, it is very difficult to control the speed particularly during the peak hours in real life. Thus in this paper, the speed is assumed to be a constant average speed, and $\varepsilon\,(v)$ will also be a constant in the paper.

3.2   MIP Model

In this section, we will describe the MIP model of the problem. Firstly, the model notations of the parameters for the problem are summarized as follows:

V: set of all vehicles.
N: set of all nodes, including the patients, the depot and the laboratory.
N': set of all nodes, including the patients, the fictive patients, the depot and the laboratory.
A': set of arcs, $A' = \{(i,j)\,|i,j \in N', i \neq j\}$.
P: set of all patients.
P': set of all patients, including the fictive patients.
Q: capacity of each caregiver.
P^{sync}: set of synchronized visits.
d_{ij}: the distance from node i to node j.
u_{ij}: the demand of patients up to node i, and transported in arc (i,j).
q_i: the demand of patient i.
τ_i: the service duration for node i.
$[a_i, b_i]$: the availability time window of patient i.

v: the speed of vehicle.
$\varepsilon(v)$: the carbon emissions function.
M: a large positive value.

In order to model the problem clearly, we adopt the most widely used three-index method to describe the MIP model. Firstly, the primary decision variable is presented as follows:

$$x_{ijk} = \begin{cases} 1, & \text{if caregiver } k \text{ travels from } i \text{ to } j, \text{ in which } i \neq j; \\ 0, & \text{otherwise.} \end{cases}$$

The secondary decision variable is denoted as follows:
y_i: the start working time of node i.
The MIP model can be denoted as follows:

$$Minimize \sum_{(i,j) \in A} \sum_{k \in V} \varepsilon(v) d_{ij} x_{ijk} \tag{3}$$

subject to,

$$\sum_{k \in V} \sum_{j \in N'} x_{ijk} = 1, \ \forall i \in P' \tag{4}$$

$$\sum_{j \in N'} x_{jik} - \sum_{j \in N'} x_{ijk} = 0, \ \forall i \in P', k \in V \tag{5}$$

$$\sum_{j \in N'} x_{0jk} \leq 1, \ \forall k \in V \tag{6}$$

$$\sum_{i \in N'} x_{i(n+1)k} \leq 1, \ \forall k \in V \tag{7}$$

$$\sum_{i \in N'} u_{ji} - \sum_{i \in N'} u_{ij} = q_j, \ \forall j \in P' \tag{8}$$

$$u_{ij} \leq Q \sum_{k \in V} x_{ijk}, \ \forall (i,j) \in A' \tag{9}$$

$$y_i - y_j + \tau_i + d_{ij}/v \leq M(1 - x_{ijk}), \ \forall i \in N', j \in P', k \in V, i \neq j \tag{10}$$

$$a_i \leq y_i \leq b_i, \ \forall i \in N' \tag{11}$$

$$y_i = y_j, \ \forall (i,j) \in P^{sync} \tag{12}$$

$$x_{ijk} \in \{0,1\}, \ \forall (i,j) \in A', k \in V \tag{13}$$

$$u_{ij} \geq 0, \ \forall (i,j) \in A' \tag{14}$$

$$y_i \geq 0, \ \forall i \in P' \tag{15}$$

The objective function (3) is the total carbon emission cost based on the speed of the vehicle, the planed distance and the carbon emissions function. Constraint (4) guarantee that each patient is visited only once. Constraint (5) ensures the flow balance of the vehicles, i.e., the caregiver visits the patient and then will leave the patient. Constraints (6) and (7) ensure that the vehicles start at the depot and end at the medical laboratory. Constraint (8) is the flow equation for the demand of patients, and constraint (9) is the capacity constraints. Constraint (10) denotes that the caregiver k can't arrive at j before $y_i+\tau_i+d_{ij}/v$, the reason is that the caregiver k needs the service duration τ_i and travel time from i to j. Here, M is a large positive value. Constraint (11) ensures the time window of the patient i. Constraint (12) guarantees the synchronized services. Constraint (13) ensures that the decision variable x_{ijk} is binary. Constraints (14) and (15) ensure the non-negative.

4   Ant Colony Optimization Algorithm

The ant colony optimization algorithm (ACO) algorithm, one of the famous swarm intelligence algorithms [10], inspired from the foraging food behavior of ant species, first proposed by Dorigo [11] for solving the traveling salesman problem (TSP), is a meta-heuristic algorithm.

4.1   Construction of Solution

In the ACO algorithm, an ant constructs a solution of the proposed problem. Each solution consists of a set of routes and each route is serviced by one vehicle. The main procedure of the ACO algorithm is presented in Algorithm 1, in which different ants obtain the information of their neighbor environment and communicate with other ants by updating the concentration of pheromone [12]. Each ant uses a probabilistic rule to select the next patient to visit based on the constraints. The probability of the ant k to visit patient j after visiting patient i is calculated as follows:

$$P_{ij}^k = \begin{cases} \frac{(\tau_{ij})^\alpha (\eta_{ij})^\beta}{\sum_{l \in C_i^k}(\tau_{il})^\alpha(\eta_{il})^\beta}, & if \ j \in C_i^k \\ 0, & otherwise \end{cases} \tag{16}$$

where τ_{ij} represents the trail of the pheromone between patient i and patient j. C_i^k denotes the set of eligible candidates that the ant k can visit after patient i. α and β are two important parameters which can determine the relative influence between the visibility and the pheromone. η_{ij} is the visibility value which is defined as follows:

$$\eta_{ij} = 1/d_{ij} \tag{17}$$

where d_{ij} denotes the distance between patient i and patient j.

In the ACO algorithm, the process of constructing a feasible solution for each ant is shown in Line 7 to Line 17 of Algorithm 1. Firstly, each ant has a list C (also named available candidates) which the patients haven't been visited; then, based on the constraints of the problem such as time windows, capacity,

Algorithm 1. Ant Colony Optimization Algorithm

Input: an instance to be solved ;
Output: the global best solution S_{best} and cost $f(S_{best})$;
 1: Initialize the parameters $n, MaxIt, \alpha, \beta, \rho, Q, MaxConst$;
 2: Initialize the pheromone matrix τ;
 3: Initialize the available candidates (C) for each ant;
 4: Set $f(S_{best}) \leftarrow inf,\ it \leftarrow 1$;
 5: **while** $it \leq MaxIt$ **do**
 6:     **for** $k \leftarrow 1, ..., n$ **do**
 7:         Choose a patient i randomly, and update C;
 8:         **while** C is not empty **do**
 9:             Calculate effective candidates C' $(C' \subseteq C)$ satisfied the constraints (capacity, time window);
10:             **if** C' is empty **then**
11:                 Ant_k returned to Lab and prepared to start from the depot;
12:                 Update C and C';
13:             **else**
14:                 Select $j \in C'$ using the equation (16);
15:                 Update C and C';
16:             **end if**
17:         **end while**
18:         **if** $f(S_{Ant_k}) - f(S_{best}) < 0$ **then**
19:             $S_{best} \leftarrow S_{Ant_k}$;
20:         **end if**
21:         $k \leftarrow k + 1$;
22:     **end for**
23:     Apply the pheromone update operator;
24:     **if** $MaxConst$ interations without improvement **then**
25:         **break while**;
26:     **end if**
27:     $it \leftarrow it + 1$;
28: **end while**

---

etc., each ant can calculate the effective candidates C' $(C' \subseteq C)$. It is clear that initial candidates are all the patients. If there is no effective candidate for an ant namely $C' = \emptyset$, the ant will return to the medical laboratory and prepare to search the patients again. If there is no available candidate for an ant namely $C = \emptyset$, the ant will stop searching, and a feasible solution will be constructed.

4.2 Pheromone Update and Stopping Criteria

The pheromone trails in the ACO algorithm are updated as follows:

$$\tau_{ij} \leftarrow (1 - \rho)\tau_{ij},\ \forall (i, j) \qquad (18)$$

where $\rho \in [0, 1]$ is an adjustable parameter of pheromone. After evaporation, the pheromones are updated as follows:

$$\tau_{ij} \leftarrow \tau_{ij} + \sum_{k=1}^{n} \Delta\tau_{ij} \qquad (19)$$

where $\Delta\tau_{ij} = \begin{cases} \frac{Q}{L_{Ant_k}}, & if \ (i,j) \in \ \text{the tour constructed by } Ant_k \\ 0, & otherwise \end{cases}$, L_{Ant_k} is the

fitness value of the objective function of the solution constructed by Ant_k, n is the numbers of all the ants, and Q is a constant.

After finishing the pheromone operator, the ACO algorithm will determine whether to terminate the iteration. If the algorithm reaches the maximum number of iterations $MaxIt$ or the algorithm doesn't obtain a better solution for $MaxConst$ iterations, the ACO algorithm will be stopped.

5   Computational Experiments

To the best of our knowledge, there are no existing benchmark instances for our HHC scheduling problem. Therefore, in order to obtain effective benchmark instances, we generate the test instances based on the classical VRPTW benchmark instances designed by Solomon in 1983. We use the proposed ACO algorithm to solve the studied problem. At the same time, the Gurobi solver is also applied to solve the studied problem, which is as a benchmark to demonstrate the effectiveness and efficiency of the proposed ACO algorithm for the studied problem.

5.1   Test Instances and Parameters Settings

There are no similar problem in the existing researches, so we generate the test instances based on the classical Solomon VRPTW benchmark instances. In the Solomon VRPTW benchmark instances, the information includes the location of the customers and depot, demand, time windows (ready time, due time), and the service time.

In the Solomon VRPTW benchmark instances, the speed is standardized to 1. It is very necessary to adjust the proportion of the data in the Solomon VRPTW benchmark instances to suit the proposed problem. According to the survey, the normal speed limit is 50 km/h in the city of France. However, the drivers often need to slow down and accelerate during driving when driving to the intersection, so it is difficult to keep an average speed at 50 km/h. In this paper, the HHC scheduling activities happens at a city or a town. Therefore, an average speed 10 m/s (namely 36 km/h) is very suitable in the test instances of the proposed problems. In the basis of the Solomon VRPTW benchmark instances, the rules of generating the test instances of the proposed problems are as follows: we set the coordinate of the medical laboratory as (30, 40); the speed is set as 36 km/h; the distance is 100 times the original, the time window and service time are 10 times the original; other parameters will not be changed. The unit of the Xcoord and Ycoord is meter(m), and the unit of the time windows and service time is second(s).

The corresponding main parameters of the paper are listed in Table 1. All the experiments are conducted on Intel Core i7-3770, 8 Duo 3.4 GHZ in order to solve the proposed problem.

Table 1. The parameters of ACO algorithm.

Algorithm	Parameters	Values
ACO	n: the number of the ants	50
	$MaxIt$: the maximum iterations	500
	α: the number of pheromones contained in a direction	1
	β: the weighting of unit quality value in a direction	2
	ρ: the evaporation rate of pheromone	0.1
	Q: a constant	1
	$MaxConst$: the maximum iterations without improvement	80

5.2   Experimental Results

As mentioned before, we generate the test instances based on the classical VRPTW benchmark instances. In the proposed MIP model, each caregiver has to begin from the depot and end at the medical laboratory, which is quite different from the classical VRPTW. The generated test instances have two important parameters, which are speed and the patient in need of synchronized service. As for the speed settings of the test instances, we set the speed as 40 km/h in the paper. As for the patient in need of synchronized service, in order to code conveniently, we use the following formulation to decide the number of synchronized-service patients:

$$NSync = \left\lceil \frac{NP}{10} \right\rceil \tag{20}$$

where $\lceil a \rceil = min\{n \in \mathbb{Z} | a \leq n\}$ which means the smallest integer larger than a, $NSync$ is the number of synchronized-service patients, and NP is the number of the patients. And we set the third patient in every ten patients as the synchronized-service patient.

In this paper, we use the proposed ACO algorithm and exact method (Gurobi solver) to solve the proposed problem. Gurobi is a pretty good commercial optimization solver, and has been used by many researchers for solving the linear programming (LP), quadratic programming (QP), quadratically constrained programming (QCP) and mixed-integer programming (MIP).

The studied problem is NP-hard, so the Gurobi solver can only solve the small scale instances within short calculating time. Therefore, we set the smallest scale instances with 10 patients. In this paper, the Gurobi solver is used with a time limit of 1 h (namely 3600 s) to perform a comparison of the performance of the proposed ACO algorithm. If the Gurobi solver doesn't give an exact solution in 3600 s, we will give the best lower bound and upper bound calculated by the Gurobi solver. We also runs the proposed ACO algorithm for 10 times, and the experimental results are presented in the following Table 2. In the tables, NP represents the number of the patients, NSync means the number of the synchronized-service patients, and the Gap is calculated as follows:

$$Gap = \frac{ACO.Best - Gurobi.Cost}{Gurobi.Cost} \times 100\% \qquad (21)$$

where ACO.Best is the best objective value calculated by the proposed ACO algorithm, and Gurobi.Cost is the result obtained by the Gurobi solver. It should

Table 2. The experimental results of transportation problems

Instance			Gurobi Solver		ACO			
Name	NP	NSync	Cost(kg)	CpuT(s)	Best(kg)	Gap(%)	Avg.	CpuT(s)
HHC_C103	10	1	[9.53,12.5]	3,600.00	12.56	0.48	12.77	3.07
HHC_C104	10	1	[8.11,12.13]	3,600.00	12.33	1.68	12.47	2.89
HHC_C105	10	1	12.50	0.63	12.50	0.00	12.50	3.30
HHC_C203	10	1	22.30	236.05	22.85	2.47	23.47	2.89
HHC_C204	10	1	21.10	680.41	21.29	0.90	21.97	3.47
HHC_C205	10	1	24.18	5.99	24.18	0.00	24.27	3.15
HHC_R104	10	1	26.52	258.99	27.59	4.05	28.34	3.20
HHC_R105	10	1	32.93	0.50	32.93	0.00	33.06	4.02
HHC_R205	10	1	25.69	1.39	25.69	0.00	27.13	2.90
HHC_RC103	10	1	24.07	1,291.29	25.11	4.31	25.81	3.11
HHC_RC104	10	1	[21.21,23.53]	3,600.00	23.81	1.17	24.94	2.97
HHC_RC105	10	1	26.77	423.20	26.83	0.23	27.19	4.36
HHC_RC203	10	1	23.02	1,647.13	23.91	3.87	24.57	2.93
HHC_RC205	10	1	24.16	593.23	24.78	2.55	25.54	2.91
HHC_C103	25	3	[26.00,32.26]	3,600.00	33.01	2.33	34.33	17.58
HHC_C104	25	3	[25.99,31.99]	3,600.00	33.58	4.97	34.93	19.46
HHC_C105	25	3	32.52	23.14	32.56	0.12	34.88	20.57
HHC_C204	25	3	[25.23,34.57]	3,600.00	36.28	4.95	45.70	19.94
HHC_R104	25	3	[43.61,60.34]	3,600.00	64.51	6.91	68.80	18.23
HHC_R105	25	3	68.99	37.08	70.27	1.86	73.59	37.43
HHC_RC105	25	3	[44.61,58.87]	3,600.00	61.35	4.21	63.12	18.86
HHC_RC204	25	3	[27.39,50.90]	3,600.00	53.02	4.16	56.68	16.43
HHC_RC205	25	3	[39.27,53.89]	3,600.00	57.08	5.92	59.14	15.02
AVG				**1791.26**		**2.48**		**9.94**
HHC_C101	100	10	–	3,600.00	144.05	–	153.13	238.29
HHC_C102	100	10	–	3,600.00	159.25	–	165.96	214.78
HHC_C103	100	10	–	3,600.00	161.24	–	175.95	156.79
HHC_R101	100	10	–	3,600.00	236.63	–	242.15	318.72
HHC_R102	100	10	–	3,600.00	205.34	–	214.32	295.35
HHC_R103	100	10	–	3,600.00	187.62	–	192.25	205.96
HHC_RC101	100	10	–	3,600.00	239.94	—	246.24	227.10
HHC_RC102	100	10	–	3,600.00	224.23	–	228.18	237.37
HHC_RC103	100	10	–	3,600.00	221.94	–	227.86	220.30

The set [a, b] represents that Gurobi doesn't give an exact solution in 3600s, a and b are the best lower bound and upper bound, respectively.

be noticed that if the Gurobi solver doesn't give an exact solution in 3600 s, we use the best upper bound as the result obtained by the Gurobi solver.

As for the small size instances with 10 and 25 patients, we can see that the Gurobi solver can't give an exact solution for most instances within the time limit of 3600 s, meanwhile the proposed can solve the studied problem within 9.94 s, which demonstrates the efficiency of the proposed ACO algorithm for small scale problems. And compared to the Gurobi solver, the gap is very small, which also highlights the effectiveness and efficiency of the proposed ACO algorithm for small scale problems. As for the large scale problems, it is clear that the Gurobi can't solve all the problems within the time limit of 3600 s, meanwhile the proposed ACO algorithm can solve the large size problems using pretty good computing efficiency, which proves the superiority of the proposed ACO algorithm for the large scale problems.

Overall, the experimental results highlight the effectiveness and efficiency of the proposed ACO algorithm. The proposed ACO algorithm calculates great results with synchronized visits and carbon emissions constraints comparing to the Gurobi solver. In addition, the proposed ACO algorithm has been evaluated on instances with 100 patients. Since almost 93% of the HHC companies can't support more than 100 patients [13], the proposed ACO algorithm can be use in the most of HHC companies in France for the planning of the logistics activities in this scale.

6   Conclusions

Transportation cost is one of the largest operating costs in HHC company daily activities, thus it is crucial to optimize daily traveling routes of the HHC vehicles in order to reduce the transportation cost meanwhile improving the service quality to patients. However, transportation has serious impacts on the environment. Therefore, it compels the managers to pay more attention to CO_2 emissions when designing the daily logistics activities. This study addresses a daily routing and scheduling problem of a HHC company with the constraints of synchronized visits and carbon emissions. We formulated the problem as a MIP model. Since the synchronized visits constraints make solving the proposed problem is more complicated than solving standard VRP related problems. In order to solve the studied problem, we developed an ant colony optimization (ACO) algorithm. The experimental results highlight the efficiency of the proposed ACO algorithm compared with the Gurobi solver with a time limit of 3600 s.

In this study, we assumed that the vehicle travels in a constant average speed, and we didn't consider traffic congestion issues. In the next work, we can set speed as a variable for the problem. In addition, we can improve the proposed ACO algorithm to solve the studied problem.

References

1. Liu, R., Xie, X., Augusto, V., Rodriguez, C.: Heuristic algorithms for a vehicle routing problem with simultaneous delivery and pickup and time windows in home health care. Eur. J. Oper. Res. **230**(3), 475–486 (2013)
2. Yuan, B., Liu, R., Jiang, Z.: Daily scheduling of caregivers with stochastic times. Int. J. Prod. Res. **56**(9), 3245–3261 (2018)
3. Liu, R., Tao, Y., Xie, X.: An adaptive large neighborhood search heuristic for the vehicle routing problem with time windows and synchronized visits. Comput. Oper. Res. **101**, 250–262 (2019)
4. Bektaş, T., Laporte, G.: The pollution-routing problem. Trasport. Res. B-Meth. **45**(8), 1232–1250 (2011)
5. Fathollahi-Fard, A.M., Hajiaghaei-Keshteli, M., Tavakkoli-Moghaddam, R.: A bi-objective green home health care routing problem. J. Clean. Prod. **200**, 423–443 (2018)
6. Xiao, L., Dridi, M., Hajjam El Hassani, A., Fei, H., Lin, W.: An improved cuckoo search for a patient transportation problem with consideration of reducing transport emissions. Sustainability **10**(3), 793 (2018)
7. Jabali, O., Van Woensel, T., De Kok, A.G.: Analysis of travel times and CO_2 emissions in time-dependent vehicle routing. Prod. Oper. Manag. **21**(6), 1060–1074 (2012)
8. Demir, E., Bektaş, T., Laporte, G.: The bi-objective pollution-routing problem. Eur. J. Oper. Res. **232**(3), 464–478 (2014)
9. Teoh, B.E., Ponnambalam, S.G., Subramanian, N.: Data driven safe vehicle routing analytics: a differential evolution algorithm to reduce CO_2 emissions and hazardous risks. Ann. Oper. Res. **270**(1–2), 515–538 (2018)
10. Wang, D., Luo, H., Grunder, O., Lin, Y., Guo, H.: Multi-step ahead electricity price forecasting using a hybrid model based on two-layer decomposition technique and BP neural network optimized by firefly algorithm. Appl. Energy **190**, 390–407 (2017)
11. Dorigo, M., Maniezzo, V., Colorni, A.: Ant system: optimization by a colony of cooperating agents. IEEE Trans. Syst. Man Cybern. B Cybern. **26**(1), 29–41 (1996)
12. Wang, X., Choi, T.M., Liu, H., Yue, X.: Novel ant colony optimization methods for simplifying solution construction in vehicle routing problems. IEEE Trans. Intell. Transp. Syst. **17**(11), 3132–3141 (2016)
13. Decerle, J., Grunder, O., Hajjam El Hassani, A., Barakat, O.: A memetic algorithm for a home health care routing and scheduling problem. Oper. Res. Health Care **16**, 59–71 (2018)

Selective Vehicle Routing Problem: A Hybrid Genetic Algorithm Approach

Andrea Posada, Juan Carlos Rivera[✉] [iD], and Juan D. Palacio [iD]

Mathematical Modeling Research Group, Universidad EAFIT, Medellín, Colombia
{aposad31,jrivera6,jpalac26}@eafit.edu.co

Abstract. In this paper we deal with a selective vehicle routing problem (SVRP), which was proposed in Posada et al. [20]. In the SVRP each node belongs to one or several clusters. Contrary to classical vehicle routing problems, here it is not necessary to visit all nodes, but to visit appropriate nodes in such a way that all clusters are visited exactly once. A genetic algorithm (GA) based on random key representation is proposed to solve this VRP variant. The proposed algorithm is a hybrid metaheuristic which integrates randomized constructive solutions, a variable neighborhood search procedure, an order-first cluster-second operator, and a mixed-integer linear model to repair unfeasible solutions. The metaheuristic is tested by using instances with up to 63 nodes adapted from the generalized vehicle routing problem (GVRP). The GVRP is a special case of this SVRP where each node belongs to exactly one cluster. The results allow to evaluate the impact of different clusters configuration on the instance complexity, the impact of each algorithm's component on the metaheuristic performance, and the efficiency of the method by a comparison with a mixed-integer linear program.

Keywords: Selective vehicle routing problem · Hybrid genetic algorithm · Combinatorial optimization · Metaheuristic algorithm

1 Introduction

The vehicle routing problem (VRP) is well-known as one of the most studied problems in the operational research and combinatorial optimization field. Given a geographically scattered set of nodes and a fleet of vehicles, the VRP aims to find a route for each vehicle in such a way that each node is visited exactly once while the total traveled distance is minimized. A large number of VRP variants have been reported in the literature including different features on vehicles, nodes and the problem network [24].

In this paper, we propose a hybrid metaheuristic algorithm for solving a variant of the VRP called *selective vehicle routing problem* (SVRP), which is described in Posada et al. [20]. As main characteristic of the SVRP, not all

Supported by Universidad EAFIT and Apolo Scientific Computing Center.

L. Idoumghar et al. (Eds.): EA 2019, LNCS 12052, pp. 148–161, 2020.
https://doi.org/10.1007/978-3-030-45715-0_12

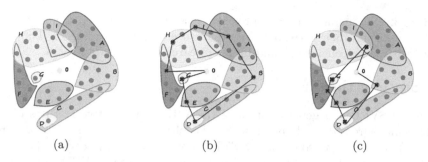

(a)   (b)   (c)

Fig. 1. Illustration of the SVRP and examples of feasible solutions. Source: Posada et al. [20].

nodes need to be visited. In contrast, in this version all of the nodes are grouped by clusters, allowing each node to belong to one or several clusters. The visited nodes must be chosen in such a way that all clusters are visited exactly once and the total traveled distance is minimized. Each node has a known demand which must be covered, by one of the available capacitated vehicles of a homogeneous fleet, when a node in the same cluster is visited. Note that the *generalized vehicle routing problem* (GVRP) is a special case of the SVRP, in which each node belongs to only one cluster. Figure 1(a) depicts an example of the SVRP studied here where 38 nodes are scattered on nine clusters represented by letters from A to H. As it is shown, the depot belongs to a single-node cluster, and different configurations can be found for required nodes: (a) single-nodes clusters, i.e. clusters with only one node (see cluster G), (b) clusters which are subsets of other larger clusters (see clusters C and D), (c) clusters without intersections with other clusters (see cluster E), and (d) clusters with non-empty intersection (see clusters A, H and I). Two examples of feasible solutions with one vehicle are shown in Fig. 1(b) and (c). Note that, both solutions visit all nine clusters, nevertheless the first one visits eight nodes while the second one visits only six.

Given the NP-hardness nature of the VRP [16], the selective variant is also an NP-hard problem. Despite the fact that exact approaches are able to solve optimally small instances of our problem [20], heuristic and metaheuristic algorithms are a suitable strategy to find good quality solutions in a decent computational time. In this paper, we deal with the SVRP via a genetic algorithm (GA). In our GA, the solutions are encoded as *random keys* [5], a kind of representation for chromosomes in GA, previously used in several combinatorial optimization problems. To evaluate and improve solution quality, we use split [22] and variable neighborhood descend (VND) algorithms. To deal with unfeasible solutions, a mixed-integer linear program (MILP) is use to repair the solution. To the best of our knowledge, there is no evidence in the literature of a similar solution strategy for selective VRPs.

The rest of the paper is organized as follows: Sect. 2 summarizes the referred literature about selective features on VRPs and metaheuristic approaches based on genetic algorithms to tackle routing problems. The SVRP is formally defined

A. Posada et al.

in Sect. 3. Our metaheuristic approach is described in Sect. 4. Computational experiments and their corresponding results are presented in Sect. 5. The paper closes with some conclusions and future research directions in Sect. 6.

2   State of the Art

Since the SVRP is a generalization of the VRP and selective features can be stated and adressed in different ways, the objective of this section is three-fold: (a) to point out some of the SVRP applications, (b) to describe briefly some of the SVRP variants reported in literature related to VRPs, i.e. GVRP, close-enough VRP (CEVRP) and clustered VRP (CluVRP). Some orienteering problems are also described: team orienteering problem (TOP), clustered orienteering problem (COP) and the set orienteering problem (SOP). Lastly, (c) to outline some of the evolutionary metaheuristic algorithms available to deal with VRPs emphasizing on genetic algorithms (GA) which is the base of the proposed hybrid metaheuristic in this paper.

Selective features for VRPs can be found in several logistic contexts. For instance, in response operations, relief aid can be delivered on close-enough nodes with the aim to cover a wider region after a disaster occurs [1]. Other applications may include: (a) the Traveling Circus Problem in which circus or exhibition managers aim to design a tour at minimum cost covering a set of villages within a maximum distance limit from each one of them, (b) material flow systems design applications for block layout of production plants as generalized routing problems, (c) urban transport routes design, (d) post-optimization procedures and subproblems of other vehicle routing models when heuristic concentration strategies, for instance, are used. For a more extensive description on selective VRPs applications, we refer the reader to Laporte et al. [15], Palekar and Laporte [18] and Baldacci et al. [4].

As mentioned in Sect. 1, one special case of the SVRP is the GVRP introduced by Ghiani and Improta [8]. The GVRP aims to design minimum cost routes starting from the depot, visiting once each one of the predefined clusters and ending each route at the depot. In the GVRP, all the clusters are mutually exclusive, that is, each node belongs only to one cluster. Despite the similarities between our problem and GVRP, ours allows to have non-empty intersections between clusters. Some of the work on the GVRP may be found in Kara and Bektaş [14], Bektaş et al. [6] and Pop et al. [19]. Secondly, the CEVRP is a variant of the SVRP in which it is not necessary to visit each node but it is possible to get close-enough to them. A well-know application of the CEVRP is the meter reading to record electricity, water and gas consumption. Utility companies with radio frequency identification technology (RFID) available are not subject to visit each one of the customers but to design close-enough routes to them. Some strategies to deal with the CEVRP may be found in Gulczynski et al. [11] and Carrabs et al. [7]. In the CluVRP, the available vehicles must visit each one of the predefined clusters but all the nodes in a cluster must be served before the vehicle leaves the cluster. We refer the reader to Sevaux and Sörensen [23] and Vidal et al. [27] for detailed algorithms dealing with the CluVRP.

Regarding orienteering problems, the TOP aims to maximize the total collected profit visiting a subset of nodes, subject to a limited route length (or time). After the orienteering problems were introduced in Tsiligirides [25] and Golden et al. [9], many scientific papers have been published. Particularly, Angelelli et al. [2] present the COP in which the maximization of total collected profit remains as objective function and a duration of the route must not exceed a given threshold. Moreover, nodes are grouped on clusters and a profit associated with each cluster is gained only if all customers in the cluster are visited. The authors in [2], propose a MILP for the COP as well as several valid inequalities. As solution strategies, they design a branch-and-cut and a tabu search and solve randomly generated COP instances. Recently, Archetti et al. [3] introduce the SOP, a generalization of the COP in which the profit of a cluster is gained if at least one of the nodes in the cluster is served. The authors propose a mathematical formulation for the problem and also a matheuristic algorithm. This matheuristic is tested on instances with up to 1000 nodes. For more details about orienting problems and related studies, we refer the reader to Gunawan et al. [12] and Vansteenwegen et al. [26].

To deal with these VRPs, many solution strategies have been used. One of them is evolutionary algorithms, particularly, GA. In basic GA, an initial population generator creates a set of randomized solutions. A selection operator selects solutions (parents) to be combined and generate new solutions (children) using a crossover operator. To avoid a premature convergence, some children are selected, usually with a small probability, to be mutated by a mutation operator. Finally, the population is updated to keep the best solutions. Depending on the tackled problem, it is possible to use different specific solution representations. Nevertheless, there exists a robust representation approach called *random keys* (RK). RK is a way to represent a solution with random numbers (or numbers in $(0, 1)$ interval) which helps to encode solutions for several combinatorial optimization problems. Precisely, Bean [5] presents a first approach to RK for GA in machine scheduling, resource allocation and quadratic assignment problems. More recently, Gonçalves and Resende [10] present a variation of RK representation in GA: the biased random-key strategy. In the biased algorithm, each new solution is generated selecting at least one parent at random from an *elite* subset of individuals. The authors present some applications for several scheduling problems (e.g., resource-constrained, multiproject, single machine).

With a more general structure, memetic algorithms (MA) are a class of metaheuristics which combine ideas from population-based algorithms and local search (LS) techniques [17]. Hart et al. [13] define MA as a population-based metaheuristic composed of an evolutionary framework and a set of local search algorithms which are activated within the generation cycle of the external framework. Nevertheless, most of the implementations in combinatorial optimization are hybrid metaheuristics combining GA with LS procedures. GA allows to explore the search space due to the fact that they maintain a pool of solutions simultaneously. On the other hand, LS procedures aim to intensify the search on promising zones of the solution space. MA has been successfully used to solve vehicle routing problems and its variants. For instance, Prins [21] proposes an MA with a split procedure to evaluate the fitness function of each individual for the CVRP and three classical heuristic procedures to generate good starting points. The algorithm is tested on instances up to 483 nodes with competitive computational times.

Vidal et al. [28] combine a GA structure with LS and population-diversity schemes with the aim to solve the multi-depot VRP, the periodic VRP and the multi-depot periodic VRP. In that algorithm, diversity and objective function values are included as criteria to preserve individuals. Authors also report new best known solutions and improvements on computational time for instances with up to 417 nodes. Vidal et al. [29] extend the work of Vidal et al. [28] by describing a population management strategy within a GA (GA|PM). In this case, the algorithm is able to deal with several VRP variants (e.g., multi-depot, periodic and vehicle-site dependencies) when time windows are imposed. This strategy is tested with instances up to 1000 nodes and 27 vehicles outperforming state-of-the-art methods.

3   Mathematical Definition

Based on Posada et al. [20], the SVRP studied in this paper can be mathematically defined as follows. Let $\mathcal{G} = (\mathcal{V}, \mathcal{E})$ be a complete graph. The set $\mathcal{V} = \{0, 1, ..., n\}$ contains the origin (depot), labeled as 0, and the set of n required nodes. The set $\mathcal{E}$ is composed by all edges or arcs (i, j), i.e. $\mathcal{E} = \{(i, j) \mid i, j \in \mathcal{V}, i \neq j\}$. Each vertex $i \in \mathcal{V}$ has associated a certain amount of demand q_i and a spatial coordinate (x_i, y_i), and each arc $(i, j) \in \mathcal{E}$ a label d_{ij}, which is the euclidean distance between nodes i and j. The set of vertices $\mathcal{V}$ is partitioned into $m + 1$ non-empty subsets, called clusters, i.e. $\mathcal{V}_0, \mathcal{V}_1, ..., \mathcal{V}_m$, such that $\bigcup_{c \in \mathcal{C}} \mathcal{V}_c = \mathcal{V}$ being $\mathcal{C} = \{0, 1, 2, ..., m\}$ the set of clusters. The cluster $\mathcal{V}_0 = \{0\}$ is composed of the depot. For modeling purposes, we define the parameter λ_{ic} which indicates whether the node $i \in \mathcal{V}$ belongs to cluster $c \in \mathcal{C}$ ($\lambda_{ic} = 1$) or not ($\lambda_{ic} = 0$). The total demand of each cluster can be satisfied if any of its nodes is visited. A fleet $\mathcal{K}$ of identical vehicles (homogeneous fleet) with individual capacity Q are based at the depot. It can be assumed, without loss of generality, that $|\mathcal{K}| \leq |\mathcal{C}|$.

A solution for the problem in $\mathcal{G}$ is a collection of $|\mathcal{K}|$ constrained routes, one for each vehicle. Each vehicle $k \in \mathcal{K}$ must perform a route which starts at the depot 0, spans a set $\mathcal{V}^k \setminus \{0\}$ of selected nodes and ends at the depot 0, while its total load cannot exceed to capacity Q. Each cluster has to be visited by one vehicle. The objective is to minimize the total traveled distance. As nodes can belong to several clusters (see Fig. 1) and each cluster must be visited, we define the parameter $w_i = \frac{1}{\sum_{c \in \mathcal{C}} \lambda_{ic}}$ for each node $i \in \mathcal{V}$, which avoids to over-count the demand of nodes when computing the demand of clusters.

As vertices are grouped by clusters, there are some special cases to take into account. A vertex can belong to several clusters, e.g. three in Fig. 1, so if it is visited, all nodes in any of its clusters are covered. That is the reason why the second solution visits less nodes. Nevertheless, the demand of nodes covered by a single node can exceed vehicle capacity if the visited node belongs to many clusters. On the other hand, a cluster can be a subset of another one (see cluster D in Fig. 1). In this case, a node from the subset cluster has to be visited, otherwise the solution is unfeasible.

Posada et al. [20] present mixed-integer linear programs (MILPs) for SVRP including alternative constraints. In Sect. 5, we compare our results with the best performed MILP in [20].

4   Metaheuristic Approach

Since VRP and GVRP are special cases of the SVRP dealt in this paper, SVRP belongs to the NP-hard class of problems and metaheuristic methods must be used to solve large instances of the selective variant. The solution strategy proposed in this paper is mainly based on GA. Initially, the procedure generates a set of feasible solutions to compose a population. In each generation, the children are created by a combination of parents chosen from the population and they are accepted in the new generation based on quality criteria. Solutions can be included in the population if they are good enough. The algorithm also uses an MILP as a repairing procedure if unfeasible solutions are obtained. The algorithm can be described by Algorithm 1.

In this section we describe the main procedures and features of the proposed metaheuristic.

Solution Representation: In our GA implementation, solutions are represented by random keys (RK). Each node has associated a priority number between 0 and 1 which allows to select the visited nodes and their order. This representation also facilitates crossover and mutation processes.

The procedure to transform a RK solution into a set of routes is the following: let RK_i be the random key number given to node i, and $\mathcal{V}_S$ and $\mathcal{V}_U$ the subsets of $\mathcal{V}$ which represent the serviced or visited and the uncovered nodes respectively. At the beginning, $\mathcal{V}_S = \emptyset$ and $\mathcal{V}_U = \mathcal{V}$. While $\mathcal{V}_U \neq \emptyset$, the node $i \in \mathcal{V}_U$ with the largest RK value is chosen to be visited. Then, for each cluster to which node i belongs to, all nodes in those clusters are removed from $\mathcal{V}_U$ and node i is added to $\mathcal{V}_S$. Nodes in $\mathcal{V}_S$ create a non-capacitated giant tour in the order they are included in the set. Lastly, $\mathcal{V}_S$ is translated into a set of routes by a split algorithm.

As initial solutions are generated with a more involved procedure, they can be easily converted to random keys. Take into account that each node in a route has a greater RK value than all nodes visited after it. Similarly, all visited nodes have greater RK values than all (not visited) nodes in the same cluster.

Generation of Feasible Solutions: This process is carried out by two randomized constructive methods. A restricted candidate list (RCL) is generated under minimum distance criterion for the first method and minimum

Algorithm 1 – GA

$POP \leftarrow$ RandomizedConstruction($nPOP$)
$POP \leftarrow$ Split(POP)
$POP_{RK} \leftarrow$ RK(POP)
while $stop\_criteria = false$ **do**
  for $i = 1$ to $nPOP$ **do**
    $Parents_{RK} \leftarrow$ Tournament(POP_{RK})
    $Son_{RK}[i] \leftarrow$ Crossover($Parents_{RK}, \alpha$)
    if rand() $< \beta$ **then**
      $Son_{RK}[i] \leftarrow$ Mutate($Son_{RK}[i]$)
    end if
    $Son[i] \leftarrow$ decodify($Son_{RK}[i]$)
    $Son[i] \leftarrow$ Split($Son[i]$)
    if infeasible($Son[i]$) **then**
      $Son[i] \leftarrow$ makeFeasible($Son[i]$)
    end if
    if rand() $< \theta$ **then**
      $Son[i] \leftarrow$ VND($Son[i]$)
    end if
  end for
  $[POP, POR_{RK}] \leftarrow$ Update($POP, Son, POP_{RK}, Son_{RK}$)
end while

density criterion for the second one. We compute the density of node j when it is visited after node i as:

$$\text{density}(i,j) = \frac{d_{ij}}{\sum_{c \in C} \lambda_{jc}} \qquad \forall\, i,j \in \mathcal{V}$$

Density criterion gives higher priority to nodes which cover a larger number of clusters and with a smaller distance to the last visited node.

Splitting Procedure: The implemented split is presented in [22]. Here, as not all nodes need to be visited, the split algorithm is only applied on the visited ones ($\mathcal{V}_S$). Let us remark that since few nodes can cover many others and a fixed size fleet is assumed, maybe not all vehicles are used. And that the split algorithm is able to provide the optimal set of routes subject to a given order (giant tour).

Selection Operator: This process is carried out by tournament. Each time, two pairs of solutions are randomly chosen; all solutions have the same probability to be selected. For each pair, the solution with less cost becomes one of the parents.

Crossover Operator: The crossover operator is based on the classical uniform method, but parents do not have necessarily the same probability; i.e. for each node, the random key is inherited from the first parent with a probability of α and from the second parent with a probability $1 - \alpha$, where $\alpha \in (0, 0.5]$. The number of generated children is equal to the population size. Note that a child solution can visit nodes that are not visited by its parents.

Improvement Operator: A variable neighborhood descend (VND) composed by 2-opt and 2-opt* moves is applied to the children with a probability θ.

Mutation Operator: The children can be also mutated with a probability β. This function operates over the RK solution. It consists of the selection of three positions of the RK solution and change their value for three new random numbers.

Population Update: After all children are generated, $nPOP$ solutions from the current and children populations are saved as the successive population for the next generation. Individuals with the best objective values over both populations have priority to be selected. This priority is proportional to the objective function value.

Repairing Procedure: The solutions obtained after the crossover and mutation operators can be unfeasible. Depending on the visited nodes, some can become unreachable or all nodes of a cluster can be covered without visiting the cluster.

Given that the number of unfeasible solutions is high, these solutions are repaired by a mixed-integer linear program. The following formulation allows to guarantee feasibility by changing previous selected nodes. The objective of this model is to preserve as many nodes as possible from the original solution. To model this MILP, let $\mathcal{P}$ be the set of nodes that cannot be visited. S_i represents the given solution where $S_i = 1$ indicates that the node i is visited. The binary decision variable x_i takes the value of one if node i is visited in the repaired solution, and of zero otherwise. Variable τ_i corresponds to the number of changes with respect to the original (unfeasible) solution. The problem can be formulated by the expressions (1)–(7).

$$\min Z = \sum_{i \in \mathcal{V}} \tau_i \tag{1}$$

$$\text{s.t.} \quad \tau_i \geq x_i - S_i, \quad \forall\, i \in \mathcal{V} \tag{2}$$

$$\tau_i \geq S_i - x_i, \quad \forall\, i \in \mathcal{V} \tag{3}$$

$$x_i = 0, \quad \forall\, i \in \mathcal{P} \tag{4}$$

$$\sum_{i \in \mathcal{V}} \lambda_{ic} \cdot x_i = 1, \quad \forall\, c \in \mathcal{C} \tag{5}$$

$$x_i \in \{0, 1\}, \quad \forall\, i \in \mathcal{V} \tag{6}$$

$$\tau_i \geq 0, \quad \forall\, i \in \mathcal{V} \tag{7}$$

Expression (1) minimizes the total number of changes to the original solution. Constraints (2) and (3) define whether there is a change over the visited nodes or not. Constraints (4) fix to zero variables corresponding to nodes that cannot be visited. Equations (5) force to visit each cluster exactly once. Finally, expressions (6) and (7) define variables domain.

Once obtained a feasible set of nodes to be visited, they are organized in a solution preserving the order of the conserved nodes and inserting the new selected ones, according to the available capacity, from highest to lowest demand.

5   Computational Experiments

The proposed metaheuristic algorithm is coded in Python 3.7 running on an Intel Xeon 2.10 GHz with 64 gigabytes of RAM running under Linux Rocks 6.2. To evaluate the performance of the proposed algorithm, we use the 48 benchmark instances solved in [20] (36 for the SVRP and 12 for the GVRP). Let us remark that SVRP instances are generated by adding some random chosen nodes on the predefined clusters of GVRP instances. SVRP instances have three levels of complexity defined as the maximum number of clusters per node (MNCN) and it may vary from two to four. Given the randomness when including nodes in each cluster, it can be noted that expression $\sum_{i \in \mathcal{V}} \sum_{c \in \mathcal{C}} \lambda_{ic}$ may also vary for each instance. Therefore, [20] propose the average cluster density (ACD) ratio as a complexity measure of each instance computed as follows:

$$ACD = \frac{|\mathcal{V}|}{\sum_{i \in \mathcal{V}} \sum_{c \in \mathcal{C}} \lambda_{ic}} \tag{8}$$

Three sets of experiments have been designed to setup and evaluate the algorithm performance. Firstly, in Sect. 5.1, we define how we generate initial solutions: selection criteria and RCL size for the randomized constructive method. Next, in Sect. 5.2, we evaluate the performance of the split procedure. Finally, Sect. 5.3 compares the results of the proposed GA and the MILP.

5.1   Initial Solutions

In this section we compare different selection criteria using deterministic constructive methods for the SVRP. Table 1 presents the results for two selection criteria based on minimum distance and minimum density.

In general, it can be observed that the distance criterion shows better results in most of the items specified in Table 1. With this criterion, the best solution is found for 19 of 36 instances of the SVRP (first row), and additionally, the same solution as with the density criterion for 5 out of 36 instances of the SVRP. Moreover, two solutions found are optimal, one of them also is met by density criterion; and another solution is equal to the one found by the mathematical model, which can not be guaranteed as optimal. The second and third row

Table 1. Comparison between selection criteria for SVRP instances

Comparison item	Distance criterion	Density criterion
Number of best solutions	19	12
Average lower bound gap	32.31%	32.83%
Average BKS gap	24.93%	25.25%
Average time (seconds)	1.95×10^{-2}	1.67×10^{-2}

Table 2. Comparison of RCL size for GVRP and SVRP instances

RCL size	GVRP					SVRP				
	gap_{avg}	No. gap_{avg}	gap_{min}	No. gap_{min}	Time (s)	gap_{avg}	No. gap_{avg}	gap_{min}	No. gap_{min}	Time (s)
1	21.14%	10	21.14%	4	4.19×10^{-3}	24.93%	15	24.93%	4	1.95×10^{-2}
2	27.85%	2	17.12%	6	4.15×10^{-3}	27.22%	8	17.55%	10	2.06×10^{-2}
3	31.89%	0	21.24%	0	4.85×10^{-3}	29.57%	4	17.10%	12	2.29×10^{-2}
4	33.55%	0	22.77%	0	5.01×10^{-3}	30.00%	3	17.46%	11	2.26×10^{-2}
5	35.65%	0	23.70%	2	4.62×10^{-3}	30.61%	3	19.69%	9	2.30×10^{-2}
6	37.31%	0	26.80%	0	4.36×10^{-3}	31.61%	0	19.54%	7	2.36×10^{-2}
7	37.90%	0	27.36%	0	4.24×10^{-3}	31.76%	0	19.42%	10	2.26×10^{-2}

show the average gap with respect to the lower bound found with CPLEX (see [20]) and with respect to the best solution found with BKS, respectively. BKS includes the solutions of the mathematical model and those found using GA. When comparing the criteria with the average gaps mentioned above (second and third row), there are some similarities in the results; however, the distance criterion is still better. For the GVRP instances, the average gap to best known solution is 21.14% when distance criterion is applied. Let us comment that in the GVRP, density and distance criteria are equivalent since $\sum_{c \in \mathcal{C}} \lambda_{jc} = 1, \forall j \in \mathcal{V}$. Given those results, minimum distance criterion is chosen for the following experiments.

Table 2 reports the results obtained on RCL size experiments for SVRP and GVRP instances, respectively. Second and third columns (gap_{avg} and No. gap_{avg}) show the average gap and the number of instances in which the minimum average gap has been found over available instances. In a similar way, fourth and fifth columns (gap_{min} and No. gap_{min}) report the average on the minimum gaps and the number of instances in which the minimum gap has been obtained, respectively. We also present the average computational time (in seconds) required to perform these set of experiments. We present experiments with values from 1 to 7 for the RCL size. Note that if RCL size increases, the average gap with respect to the best known solution also increases. Lower RCL size also gets more times the minimum average gap for both type of problems. However, with respect to the minimum gap found, the behavior is not the same. The minimum gaps are found when the RCL size is between 3 and 4 for SVRP. For GVRP, an RCL size equal to 2 has a better performance. There is not significant difference between computing times for different RCL sizes. Nevertheless, the algorithm is faster solving GVRP instances. In order to establish the initial population with diverse solutions we use the seven RCL size values to build solutions. Each iteration the size of the RCL is randomly chosen.

5.2 Performance of Split Procedure

As it has been noticed before, given an initial solution, the split procedure can find a new solution with the same or better quality. For the SVRP, the split improves quality on 68.82% of the solutions and therefore, finds the same quality for 31.18% of the population. The average gap improvement per instance is 2.85%. On the other hand, for the GVRP, the algorithm has a lower success: it improves quality on 28.62% of the individuals and then, 71.38% of them, remains the same. The average gap improvement per instance is 2.03%. Although, the procedure is efficient and effective for both type of problems, but it is 35% faster for SVRP than for GVRP. The average computing time is 3.18×10^{-4} s and 4.94×10^{-4} s for the SVRP and GVRP, respectively.

5.3 Genetic Algorithm Performance

In order to tune up the algorithm, four parameters have been evaluated with the following values: population size $\in \{50, 100\}$, $\alpha \in \{0.3, 0.5\}$, $\beta \in \{0.01, 0.05\}$ and $\theta \in \{0, 0.05, 0.1\}$. Each of the 24 combinations have been run for all instances with 10 min as stop criterion. Although there is not a large difference, the best results have been obtained with 100 individuals, $\alpha = 0.5$, $\beta = 0.01$ and $\theta = 0.1$. On SVRP instances the algorithm finds 21 out of 36 optimal solutions and the average gap between the best solution found and the best lower bound gets by CPLEX [20] is 11.73%. On GVRP instances the algorithm finds 4 out of 12 optimal solutions and the average gap is 22.66%. We also noticed that on average the best solution found by GA is obtained in the first 9.8% (59.0 s) of the running time for SVRP instances; and 39.7% (238.4 s) of the running time for GVRP instances. In the following tables we use 10 min as stop criterion but we reported average running time per iteration; that helps us to approximate the instance complexity since more complex instance use more frequently the repairing procedure and take more time for each iteration.

Tables 3 and 4 summarize the GA results. Table 3 classifies the instances with different ACD values. The first row presents the number of instances on each category. Second one shows the number and percentage of optimal solutions found. Third and forth rows indicate the average gap with respect the best known solution and the lower bound found by CPLEX, respectively. Finally, fifth row displays the average running time per iteration in seconds. It can be seen that there is a relationship between instance complexity and the average cluster density (ACD). Instances with higher ACD value gets larger gaps (BKS and LB) and running times, and less number of optimal solutions. For last two columns there is no a significant difference on the number of optimal solutions.

Similarly, Table 4 summarizes the GA results where instances are classified by MNCM (maximum number of cluster per node). Table 4 uses the same rows than Table 3. In this case, there is not a clear relationship between complexity or algorithm performance and the MNCN value.

The best performance is obtained for instances with MNCN = 2: GA reaches a higher number of optimal solutions, the best average gaps between found solutions and best known solutions and lower bounds of CPLEX, and the least average running time per iteration. The second best results are found for instances with MNCN = 4.

Finally, the results obtained by the proposed genetic algorithm are compared with the ones from the MILP presented in [20]. Table 5 presents the results for

Table 3. Performance comparison of GA for different ACD ratio values

	ACD < 0.5	$0.5 \leq$ ACD ≤ 0.65	ACD > 0.65
Number of instances	15	12	9
Optimal solutions	13 (86.7%)	5 (41.7%)	4 (44.4%)
Average gap BKS	1.17%	3.43%	3.74%
Average gap LB	2.96%	18.67%	21.05%
Average time per iteration (s)	0.46	0.89	1.04

Table 4. GA performance comparison for different MNCN values

	MNCN = 1	MNCN = 2	MNCN = 3	MNCN = 4
Number of instances	12	12	12	12
Optimal solutions	4 (33.3%)	10 (83.3%)	5 (41.7%)	7 (58.3%)
Average gap BKS	4.75%	1.42%	4.22%	2.05%
Average gap LB	24.91%	3.67%	21.87%	12.62%
Average time per iteration (s)	3.55	0.51	0.97	0.77

each method using the MNCN classification. For each method and each instance category, two results are shown: the average gap with respect to the best known solution (Gap_{bks}) and the average running time in seconds (Time). For GA, the average running time is the stop criterion. It can be noticed that, on average, GA finds closer solutions to the best known solutions, only overcame by MILP on instances with MNCN = 2 when MILP gets all best known solutions. Running time of GA is about 22% of the MILP running time, which represents the main advantage of metaheuristic methods. Nevertheless, as it have been mentioned before, stop criterion can be decreased since most of the experiments does not report improvements after 2 min.

Table 5. Performance comparison of GA and MILP for different MNCN values

Method	GVRP				SVRP			
	MNCN = 1		MNCN = 2		MNCN = 3		MNCN = 4	
	Gap_{bks}	Time	Gap_{bks}	Time	Gap_{bks}	Time	Gap_{bks}	Time
GA	0.14%	600	0.82%	600	0.19%	600	0.12%	600
MILP	0.65%	5 300	0.00%	817	1.27%	4 480	0.38%	301

6   Conclusions

The difficulty in the solution of the SVRP is proved by the need to use an additional procedure to make the solutions feasible due to the high number of non-feasible solutions. This procedure is proposed as an MILP, which shows satisfactory results.

Based on the experiments carried out, it is possible to conclude the existence of a positive relation between the ACD value and the instance complexity. That means that for lower ACD values, the required time to find a solution is also lower and the obtained solution has a higher quality. In opposition, high ACD values require more time and gets less quality on average. That is not the case for MNCN classification where no relationship with complexity were found.

The proposed GA shows good results when compared with ones from the exact methods, presented in [20]. These good results are reflected in a significant reduction of the computational time and solutions very close to the best ones.

As future work we suggest the implementation of more involved procedures that help to improve solution quality for the more complex instances. The presence of clusters based on a geographical location or more scattered ones can have different effects on the solution and on applicable methods. In addition, it is important to work on real cases where clusters presents a similar structure.

References

1. Afsar, H.M., Prins, C., Santos, A.C.: Exact and heuristic algorithms for solving the generalized vehicle routing problem with flexible fleet size. Int. Trans. Oper. Res. **21**(1), 153–175 (2014)
2. Angelelli, E., Archetti, C., Vindigni, M.: The clustered orienteering problem. Eur. J. Oper. Res. **238**(2), 404–414 (2014)
3. Archetti, C., Carrabs, F., Cerulli, R.: The set orienteering problem. Eur. J. Oper. Res. **267**(1), 264–272 (2018)
4. Baldacci, R., Bartolini, E., Laporte, G.: Some applications of the generalized vehicle routing problem. J. Oper. Res. Soc. **61**(7), 1072–1077 (2010). https://doi.org/10.1057/jors.2009.51
5. Bean, J.C.: Genetic algorithms and random keys for sequencing and optimization. ORSA J. Comput. **6**(2), 154–160 (1994)
6. Bektaş, T., Erdoağn, G., Røpke, S.: Formulations and branch-and-cut algorithms for the generalized vehicle routing problem. Transp. Sci. **45**(3), 299–316 (2011)
7. Carrabs, F., Cerrone, C., Cerulli, R., Gaudioso, M.: A novel discretization scheme for the close enough traveling salesman problem. Comput. Oper. Res. **78**, 163–171 (2017)
8. Ghiani, G., Improta, G.: An efficient transformation of the generalized vehicle routing problem. Eur. J. Oper. Res. **122**(1), 11–17 (2000)
9. Golden, B.L., Levy, L., Vohra, R.: The orienteering problem. Naval Res. Logist. **34**(3), 307–318 (1987)
10. Gonçalves, J.F., Resende, M.G.: Biased random-key genetic algorithms for combinatorial optimization. J. Heuristics **17**(5), 487–525 (2011). https://doi.org/10.1007/s10732-010-9143-1
11. Gulczynski, D.J., Heath, J.W., Price, C.C.: The close enough traveling salesman problem: a discussion of several heuristics. In: Alt, F.B., Fu, M.C., Golden, B.L. (eds.) Perspectives in Operations Research. Operations Research/Computer Science Interfaces Series, vol. 36, pp. 271–283. Springer, Boston (2006). https://doi.org/10.1007/978-0-387-39934-8_16

12. Gunawan, A., Lau, H.C., Vansteenwegen, P.: Orienteering problem: a survey of recent variants, solution approaches and applications. Eur. J. Oper. Res. **255**(2), 315–332 (2016)
13. Hart, W., Krasnogor, N., Smith, J.: Memetic evolutionary algorithms. In: Hart, W.E., Smith, J., Krasnogor, N. (eds.) Recent Advances in Memetic Algorithms. Studies in Fuzziness and Soft Computing, vol. 166, pp. 3–27. Springer, Heidelberg (2005). https://doi.org/10.1007/3-540-32363-5_1
14. Kara, I., Bektaş, T.: Integer linear programming formulation of the generalized vehicle routing problem. In: Proceedings of the 5th EURO/INFORMS Joint International Meeting (2003)
15. Laporte, G., Asef-Vaziri, A., Sriskandarajah, C.: Some applications of the generalized travelling salesman problem. J. Oper. Res. Soc. **47**(12), 1461–1467 (1996). https://doi.org/10.1057/jors.1996.190
16. Lenstra, J.K., Kan, A.R.: Complexity of vehicle routing and scheduling problems. Networks **11**(2), 221–227 (1981)
17. Neri, F., Cotta, C.: Memetic algorithms and memetic computing optimization: a literature review. Swarm Evol. Comput. **2**, 1–14 (2012)
18. Palekar, U., Laporte, G.: Some applications of the clustered travelling salesman problem. J. Oper. Res. Soc. **53**(9), 972–976 (2002). https://doi.org/10.1057/palgrave.jors.2601420
19. Pop, P.C., Kara, I., Marc, A.H.: New mathematical models of the generalized vehicle routing problem and extensions. Appl. Math. Model. **36**(1), 97–107 (2012)
20. Posada, A., Rivera, J.C., Palacio, J.D.: A mixed-integer linear programming model for a selective vehicle routing problem. In: Figueroa-García, J.C., Villegas, J.G., Orozco-Arroyave, J.R., Maya Duque, P.A. (eds.) WEA 2018. CCIS, vol. 916, pp. 108–119. Springer, Cham (2018). https://doi.org/10.1007/978-3-030-00353-1_10
21. Prins, C.: A simple and effective evolutionary algorithm for the vehicle routing problem. Comput. Oper. Res. **31**(12), 1985–2002 (2004)
22. Prins, C., Lacomme, P., Prodhon, C.: Order-first split-second methods for vehicle routing problems: a review. Transp. Res. Part C: Emerg. Technol. **40**, 179–200 (2014)
23. Sevaux, M., Sörensen, K., et al.: Hamiltonian paths in large clustered routing problems. In: Proceedings of the EU/MEeting 2008 Workshop on Metaheuristics for Logistics and Vehicle Routing, EU/ME, vol. 8, pp. 411–417 (2008)
24. Toth, P., Vigo, D.: An overview of vehicle routing problems. In: The Vehicle Routing Problem, pp. 1–26. SIAM (2002)
25. Tsiligirides, T.: Heuristic methods applied to orienteering. J. Oper. Res. Soc. **35**(9), 797–809 (1984)
26. Vansteenwegen, P., Souffriau, W., Van Oudheusden, D.: The orienteering problem: a survey. Eur. J. Oper. Res. **209**(1), 1–10 (2011)
27. Vidal, T., Battarra, M., Subramanian, A., Erdogan, G.: Hybrid metaheuristics for the clustered vehicle routing problem. Comput. Oper. Res. **58**, 87–99 (2015)
28. Vidal, T., Crainic, T.G., Gendreau, M., Lahrichi, N., Rei, W.: A hybrid genetic algorithm for multidepot and periodic vehicle routing problems. Oper. Res. **60**(3), 611–624 (2012)
29. Vidal, T., Crainic, T.G., Gendreau, M., Prins, C.: A hybrid genetic algorithm with adaptive diversity management for a large class of vehicle routing problems with time-windows. Comput. Oper. Res. **40**(1), 475–489 (2013)

Fixed Jobs Multi-agent Scheduling Problem with Renewable Resources

Boukhalfa Zahout, Ameur Soukhal[✉], and Patrick Martineau

LIFAT EA 6300, CNRS, ROOT ERL-CNRS 7002, Université de Tours, Tours, France
{boukhalfa.zahout,ameur.soukhal,patrick.martineau}@univ-tours.fr

Abstract. We consider multi-agent scheduling problem where the set of jobs to schedule is divided into two disjoint subsets A and B. Each subset of jobs is associated to one agent. The two agents compete to perform their independent jobs without preemption on common m identical parallel machines. Each machine has limited renewable resource units $R_k, k = 1 \ldots K$ necessary to perform each job. The start date, fixed finish date and required additional resources are given and fixed. A machine can process more than one job at a time provided the resource consumption does not exceed R_k. The objective is to determine a feasible schedule that maximizes the number of scheduled jobs of agent A, while keeping the number of scheduled jobs of agent B no less than a fixed value Q_B, or equivalently the agents aims at minimizing the number of their rejected jobs. This problem is called a *Competing multi-agent scheduling*. The problem under study is $\mathcal{NP}$-hard. To obtain best compromise solutions for each agent, integer linear programming model and greedy heuristics based on ε-constraint approach are proposed to compute exact and approximate Pareto fronts. A *Non-dominated Sorting Genetic Algorithm (NSGA-II)* is developed to generate Pareto front. Experimental results are driven to analyse the performances of the proposed methods.

Keywords: Competing multi-agent scheduling · Fixed job scheduling · Resource allocation · Parallel machines · Linear integer program · Greedy heuristics · NSGA-II · Pareto fronts

1 Introduction

Scheduling problems are an important part of combinatorial optimization problems. They are encountered in any operating system when it comes to organize activities or tasks over time and determine their best allocation(s) to consumable or renewable resources. The successful management of large-scale job processing systems is a difficult problem, especially in the presence of multiple users.

In real situations, the jobs share common resources to be scheduled where different objective functions should be optimized. We face therefore multi-criteria scheduling problems. Classical multi-criteria scheduling problem have been widely studied in the literature (see [18] and [11]) where it is assumed that the quality of solution is measured by one or more objective functions applied

© Springer Nature Switzerland AG 2020
L. Idoumghar et al. (Eds.): EA 2019, LNCS 12052, pp. 162–176, 2020.
https://doi.org/10.1007/978-3-030-45715-0_13

to all jobs without distinction. In some real situations, this assumption may not hold or it may not be appropriate to the more general resource allocation problem. Instead of using one or more criteria for all jobs, we may need to consider that each objective function depends on a subset of jobs. For instance, it is possible to consider a workshop where some jobs may have a soft due date with allowed tardiness (to be minimized); whereas some other jobs may have hard due dates (that must be respected) and other jobs may have no due date (minimize the dwell) [19]. The dwell is commonly expressed in the scheduling literature by the total completion times. This is a multi-criteria scheduling problem where a new type of compromise must be achieved. These scheduling problems are referred to in the literature as "multi-agent scheduling" [7,13] or "scheduling with competing agents" [3] or "interfering job scheduling problems" [11,19].

A scheduling problem involving several actors, where each has its own decision-making autonomy, in charge of executing its subset of jobs on the same resources (the jobs are competing for the use of the same machines), can be assimilated to a multi-agent scheduling problem, where a new type of compromise must be achieved. We define the term "agent" as an entity associated with a subset of jobs. This entity may be associated with another decision-maker who intervenes in the choice of the final solution. Each agent aims at minimizing a criterion of his own because it depends only on his jobs. These agents compete since they share the same resources [1,2]. We are therefore looking for well compromised solutions. These problems are close to the combinatorial optimization area and cooperative game theory [4].

In this paper, we are mainly interested in competing multi-agent scheduling problem. The rest of this paper is organized as follows. Section 2 introduces the problem definition and notations. A literature survey on related problems is presented in Sect. 3. In Sect. 4 we propose exact method based on time-indexed variables to compute an exact Pareto front. Section 5 introduces two greedy heuristics where ε-constraint approach is used to compute approximate Pareto front. A non-dominated sorting genetic algorithm (NSGA-II) is developed in Sect. 6. A discussion on the performance of the proposed methods is provided in Sect. 7. Section 8 concludes and provides some future research directions.

2 Problem Definitions

To illustrate our approaches, we focus on the case of two agents A and B. All the results developed in this paper can be generalized to L agents. Agent A (resp. B) is associated with the set of n_A (resp. n_B) jobs, denoted by $\mathcal{N}^A = \{J_1, J_2, ..., J_{n_A}\}$ (resp. $\mathcal{N}^B = \{J_{n_A+1}, J_{n_A+2}, ..., J_n\}$), where $n = n_A + n_B$.

The n independent jobs should be scheduled without preemption on m identical parallel machines. Additional renewable resources are however necessary to process each job. Several types of such resources, denoted $R_k, k = 1 ... K$, are needed. Hence, at execution time of job j, r_{jk} units of resource R_k are required. For each job j, the start date s_j and its finished date f_j $(j = 1, \ldots, n)$ are fixed where its processing time $p_j = f_j - s_j$. Dealing with each type of resources k, the machine can process more than one job at a time provided the resource consumption does not exceed a given value R_k $(k = 1 \ldots K)$. We assume that the machines are continuously available during the time interval $[0, \infty)$. All data are assumed positive integers. The processing times of jobs are formatted in slotted windows. The total time period $[T_{min}, T_{max}]$ is partitioned into equal length slots (l_0) with $T_{min} = min_{j,j=1,\ldots,n}(s_j)$ and $T_{max} = max_{j,j=1,\ldots,n}(f_j)$. We suppose that: $s_j < f_j$ and $r_{jk} \leq R_k$ for all $j = 1, \ldots, n$ and $k = 1, \ldots, K$. The objective of each agent is to maximize the number of jobs that can be scheduled (or equivalently the minimize the number of rejected jobs). Let x_{ij} be the binary decision variable where $x_{ij} = 1$ if machine i processes job J_j; 0 otherwise. We denote the maximum number of jobs of Agent A and of agent B that can be scheduled by $Z^A = \sum_{i=1}^{m} \sum_{j=1}^{n_A} x_{ij}$ and $Z^B = \sum_{i=1}^{m} \sum_{j=n_A+1}^{n} x_{ij}$, respectively. In this study, ε-constraint approach is used to determine one Pareto optimal solution and also to compute the Pareto front.

According to the three-field notation of multiagent scheduling problems introduced in [1], the problems we address are denoted by:

- $Pm|CO, s_j, f_j, r_{jk}, Q_B|\varepsilon(Z^A/Z^B)$: maximize Z^A such that $Z^B \geq Q_B$
- $Pm|CO, s_j, f_j, r_{jk}|\mathcal{P}(Z^A, Z^B)$: compute the Pareto front.

These problems are $\mathcal{NP}$-hard even if only one agent is considered (mono-criterion case) [21].

The studied problem in this paper can be met in a Data center for example, where the goal is to optimize the objective function of each user (agent). Jobs (containers) submitted by the users should be executed on the different clusters by a virtualization software (Docker, for example). The clusters owns certain limited types of renewable resources CPU, MEMORY and STORAGE, with capacities Qu_1 of CPU, a certain quantity of memory Qu_2 and a certain storage capacity Qu_3. In this case, to execute $Container_j$, a number of virtual CPUs r_{j1}, virtual memory r_{j2} and hard drives r_{j3} are needed. Let us consider an example with $n = 8$ jobs of 2 agents A and B. The jobs have to be scheduled on 2 machines. Agent A's (resp. Agent B's) jobs are $\mathcal{N}^A = \{J_1, J_2, ..., J_4\}$ (resp. $\mathcal{N}^B = \{J_5, J_6, ..., J_8\}$). 3 types of resources, $R_{CPU} = R_{MEMORY} = R_{STORAGE} = 1000$ are needed. For each job j, the starting date, the finishing date and the quantities of each requirement resource $r_{jk}; k = 1, 2, 3$ are given in Table 1. Figure 1 shows an example with a strict Pareto solution.

3   Related Work

To the best of our knowledge, Peha was the first author who provided polynomial time algorithms for particular multi-agent scheduling problems dealing with n jobs and m identical parallel machines [15]. He considered an integrated-services packet-switched networks such as ATM (Asynchronous Transfer Mode). The information carried by the network are first split into smaller messages called *packets*. The data comes from different types, such as voice, video, image and

Table 1. Instance with 8 jobs and 3 types of resources.

Jobs	s_j	f_j	CPU	MEM	STORAGE
1	0	4	500	300	400
2	0	6	125	500	250
3	1	4	125	600	300
4	1	5	250	800	600
5	3	6	500	700	250
6	3	8	500	300	700
7	4	8	125	250	200
8	5	9	250	500	300

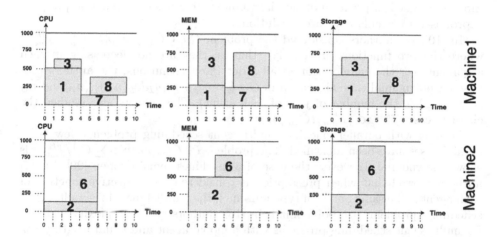

Fig. 1. Example with 4 jobs for each agent: a feasible solution.

so on. Each packet is wrapped with the essential information needed to get it from its source to the correct destination. In the case of audio and video data the author shows that minimizing the number of late delivered packets is more relevant whereas for the other types of data, minimizing the dwell is more suitable.

Since 2000, the study of multi-agent scheduling problems has grown significantly due to their practical and theoretical benefits, highlighted in [1]. In 2000 Agnetis et al. [2] were the first to introduce definitions and concepts of multi-agent scheduling problems. In their study the ε-constraint approach is considered to minimize regular criteria. New complexity results and dynamic programming algorithms have been developed.

In [1], a large and complete state of the art is established. The authors give an introduction to multi-agent scheduling, introducing general definitions and notations. Dealing with several agents the authors identified four different scenarios, in particular the Competing scenario denoted "CO". However, few results are dedicated to the case of m parallel machines. Note that problems $Pm|CO|f^A, f^B$ are $\mathcal{N}P$-hard, whatever the considered classical regular objective functions.

Balasubramanian et al. [6] have studied the case of identical parallel machines with two agents, denoted $Pm|CO|\varepsilon(\sum C_j^A / C_{max}^B)$. The problem is binary $\mathcal{N}P$-hard. The authors developed efficient heuristics to enumerate the Pareto front. As it is the C_{max}^B criterion, on each machine the jobs of agent B are grouped into one block, thus separating agent A's jobs into two blocks. The jobs of agent A (resp. agent B) are scheduled according to the SPT (resp. LPT) order. An evolutionary algorithm has been developed to compute an approximate Pareto front of the problem. The authors also propose a ILP to generate all the strictly non-dominated solutions in an iterative way. In their study, the ε-constraint approach is considered. Similarly, an optimal solution for the $Pm||\varepsilon(\sum w_j C_j^A / C_{max}^B)$ can be obtained in pseudo-polynomial running time where a dynamic program is proposed to calculate a Pareto solution.

In [10], the authors considered the problem $Qm|CO, p_j = p|\mathcal{P}(f_{max}^A, f_{max}^B)$ where the two functions are regular, that is, they are nondecreasing in each argument. For the enumeration of all strict Pareto solutions, the authors proposed a polynomial time algorithm in $O(n_A^2 + n_B^2 + n_A n_B log(n_B))$ (n_A and n_B correspond to the number of jobs of each agent). The authors also studied the classical cases $f_{max}^k \in \{L_{max}^k, C_{max}^k\}$.

Dealing with parallel machines multi-agent scheduling problems, few polynomial cases have been identified. The problem $P2|CO, pmtn|\varepsilon(\sum C_j^A / f_{max}^B)$ is shown polynomial. However, the case of 3 machines remains open [20]. In [17] authors showed that when preemption is considered and objective function of each agent is the same and is of type min-max, the parallel machines multi-agent scheduling problems are polynomial. In [16], authors gave complexity analyses for multi-agent scheduling problems with a global agent and equal length jobs.

Dealing with Data center to minimize the number of rejected jobs has been addressed in [5] where the authors have considered only one additional resource (memory) and develop methods to determine if a feasible schedule exist for all jobs. In this study, all jobs belong to a single user (only one agent is considered).

In the context of scheduling problems in grid computing, in [8] the authors considered organizations that share clusters to distribute peak workloads among all the participants. Each cluster is associated with one agent where the global objective function is introduced to minimize the makespan. The authors propose a 2-approximation algorithm for finding collaborative solutions.

4   Integer Programming Formulation

We present in this section the following time indexed integer linear programming (ILP). Let x_{ij} be a binary variable equal to 1 if machine i processes job J_j; 0 otherwise. And let y_{ijt} be a binary variable equal to 1 if machine i processes job J_j at time t; 0 otherwise.

The general formulation of the time-indexed ILP, is the following.

Maximize: $\displaystyle\sum_{i=1}^{m}\sum_{j=1}^{n_A} x_{ij}$

subject to: $\displaystyle\sum_{t=s_j}^{f_j-1} y_{ijt} = (f_j - s_j) * x_{ij}$ $i = 1, \dots, m;\ j = 1, \dots, n$ (1)

$\displaystyle\sum_{j=1}^{n_A} y_{ijt} * r_{jk} \le R_k$ $i = 1, \dots, m;\ k = 1, \dots, K; \forall t \in [T_{min}, T_{max}]$ (2)

$\displaystyle\sum_{i=1}^{m} x_{ij} \le 1$ $j = 1, \dots, n$ (3)

$\displaystyle\sum_{i=1}^{m}\sum_{j=1}^{n_B} x_{ij} \le Q_B$ (4)

$x_{ij} \in \{0,1\},\ y_{ijt} \in \{0,1\},\ i = 1, \dots, m,\ j = 1, \dots, n,\ \forall t \in [T_{min}, T_{max}]$

The constraints (1) ensure that if job J_j is not rejected then it is scheduled during its time interval. The constraints (2) ensure that no more than R_k quantities of the required resources are consumed at time t. The constraints (3) allow job j to be assigned to at most one machine. The constraint (4) express the ε-approach bounds.

In Table 2, we specify the criterion to be optimized and all of the variables and constraints taken into account.

Table 2. ILP model.

Problem	Z	# bin. var.	# constraints
$Pm\vert CO, s_j, f_j, r_{jk}, Q_B\vert \varepsilon(Z^A/Z^B)$	$\displaystyle\sum_{i=1}^{m}\sum_{j=1}^{n_A} x_{ij}$	$mn(1+T)$	$mn + mKT + n + 1$

$T = T_{max} - T_{min}$

4.1 ε-constraint Approach

To solve the ε-constraint version of the studied problem, we propose to use the previous ILP, in which the objective function of agent B is bounded. To generate the set of strict Pareto solutions, we solve $Pm\vert CO, s_j, f_j, r_{jk}, Z^B \le Q_B\vert Z^A$ with different values of $Q_B \in \{0, \dots, n_B\}$. In first step, Q_B is fixed to 0 that is the lower bound of Z^B. The obtained solution is denoted $(\hat{Z}^A, \hat{Z}^B)$. We then solve the inverse problem $Pm\vert CO, s_j, f_j, r_{jk}, Z^A \le \hat{Z}^A\vert Z^B$. The obtained solution is then optimal Pareto solution, denoted by $(\hat{Z}^A, \hat{Z}^{B\prime})$, and added to the set of strict solutions $\mathcal{S}$. We then set $Q_B = \hat{Z}^{B\prime} + 1$ and iterate. If no feasible solution is obtained then stop. $\mathcal{S}$ is then the exact Pareto front.

4.2 Number of Pareto Solutions

In this section, we indicate the number of strict Pareto solutions for the studied problem $Pm|CO, s_j, f_j, r_{jk}|\mathcal{P}(Z^A, Z^B)$. It is clear that this problem admits a polynomial number of Pareto solutions bounded by $min(n_A, n_B) + 1$. Without loss of generality since the problem is symmetrical, let us suppose that $n_B = min(n_A, n_B)$. It means that in the worst case scenario, all of Agent B's jobs are rejected, which allows for the maximum scheduling of Agent A's jobs. As well, starting from $Q_B = 0$ and iteratively, the procedure stops when $Q_B = n_B$.

5 Greedy Heuristics

The scheduling problems under study are oftentimes applied in settings where the number of jobs to be processed can be extremely large. Thus the need for low computational running time. In other words, the algorithms must have a low complexity (not just polynomial). In this section, we present low-complexity ($O(n \log n)$) greedy algorithms. Roughly, this algorithm works as follows. ε-constraint approach is used, jobs of each agent are sorted according to a given priority rule. At first, we try to schedule jobs of agent B with respect of its objective (i.e. $Z^B \geq Q_B$). Jobs are taken according to their priority order. Job is rejected if it can not be scheduled. Then, within the obtained solution, we try to schedule jobs of agent A to maximize Z^A. After testing several known priority rules, we short-listed those that offer the best performance to solve the studied problem, described as follows.

1. **Shortest Processing Time First** (*SPT*): The jobs are sorted in non-decreasing order of $(f_j - s_j)$, in case of ties, the job with the smallest finishing time comes first, otherwise a lexicographical order is considered. This *SPT* rule allows resources to be released as soon as possible.
2. **Capacity-Makespan** (*CM*): The jobs are sorted in a non-decreasing order of their occupied space given by the following formula: $(\sum_{k \in R} r_{jk} * (f_j - s_j))$, in case of ties, the job with the smallest finishing date comes first, otherwise a lexicographical order is considered. The idea of using *CM* rule is to minimize the space occupied by jobs defined by processing time per quantities of consumed resources.

According to the order obtained by the priority rules, we assign each job to the first available machine (FAM).

6 NSGA-II Algorithm

We propose a non-dominated sorting genetic algorithm called NSGA-II [9]. This method is the commonly-used evolutionary algorithm for solving multiobjective optimization problems. Based on a genetic algorithm, NSGA-II uses a ranking selection method to emphasize current non-dominated solutions and a niching method to maintain diversity in the population. We refer to [12] for a tutorial on multiple-objective optimization methods using genetic algorithms.

6.1   Encoding Mechanisms

Given the assignments of jobs, we can compute the objective function values of a non-dominated solution for the $Pm|CO, s_j, f_j, r_{jk}|\mathcal{P}(Z^A, Z^B)$ scheduling problem. Therefore, the proposed encoding is based on a *machine-assignment* scheme. An individual is a n-vector where each element j stores the number of the machine that performs job J_j. If job j is rejected, element j stores number 0. Recall previous example with $m = 2$ machines, $n = 8$ jobs and $n_B = 4$. We remember that the first four jobs belong to $\mathcal{N}^A$. Figure 2 shows the encoding and decoding of a solution.

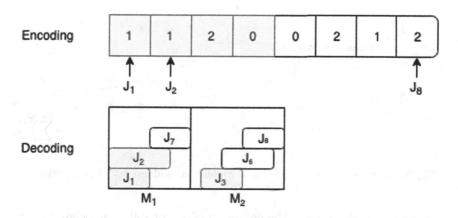

Fig. 2. Encoding and decoding mechanism.

6.2   Generation of the Initial Population

There are several ways to generate an initial population for an evolutionary algorithm: random generation, local search methods, metaheuristics, etc. We often choose a method that guarantees the diversity of the initial population. Here, the initial population is constructed using random generation and using the two heuristics presented in the Sect. 5. Let $N_{\mathcal{P}^0}$ be the size of population $\mathcal{P}^0$. This size is set at $N_{\mathcal{P}^0} = n + n/2$ which gives the best compromise between the quality of solutions and the computation time.

Dealing with criteria space Fig. 3 shows some examples of populations generated using the techniques discussed previously.

6.3   Non-dominated Sorting and Ranking

In this step, the population of individuals is classified into successive non dominated ranks or fronts, called rank 1, rank 2, and so on. This means that the solutions belonging to rank q are non-dominated among themselves and are dominated by the solutions of rank $k - 1$. We use ranking procedure proposed in [14] which is based on binary search.

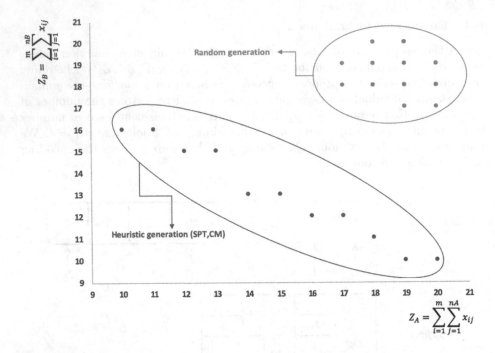

Fig. 3. Instance with $n_A = 20$, $n_B = 20$ and 2 machines: initial population.

The ranks are also used in the definition of a probability associated with an individual. The best individuals have a rank equal to 1, and we would like to attribute to them the highest possible probability. The idea is to promote the best individuals for the improvement of the next populations. Let Ra_λ be the rank of individual λ and $Ra_{\max}$ the maximum rank. The expression of probability Pr_λ, of $\lambda \in \mathcal{P}$ is the following: $Pr_\lambda = \frac{Ra_{\max} - Ra_\lambda}{N_\mathcal{P} Ra_{max} - \sum_{p \in \mathcal{P}} Ra_p}$

We note that $\sum_{\lambda \in \mathcal{P}} Pr_\lambda = 1$.

6.4 Selection, Crossover and Mutation

Let N_{C^q} be the size of the set of individuals C^q generated by the crossover operator, let N_{M^q} be the size of the set of individuals M^q generated by the mutation operator and let N_{p^q} be the size of population $\mathcal{P}^q$. To obtain a child solution c_1, eight individuals are selected randomly from the current population $\mathcal{P}^{q-1}$, two parents λ_1, λ_2 are chosen by tournament selection. A value σ is randomly generated between 0 and 1. If $\sigma \leq \sigma_{cross}$, λ_1, λ_2 are crossed and the two offsprings are added to population C^q. A value ρ is randomly generated between 0 and 1. If $\rho \leq \rho_{mutation}$, child c_1 mutates and the obtained child c_1' is added to population M^q.

We explain now the proposed crossover operator dealing with two selected individuals as illustrated in Fig. 4. The first (resp. second) child c_1 (resp. c_2) resulting from the crossing of both parents λ_1 and λ_2 is obtained as follows:

let r be the number of genes that c_1 (resp. c_2) will inherit from λ_2 (resp. λ_1), where r is randomly generated in $[1, \frac{4n-4}{n}]$; c_1 (resp. c_2) therefore inherits the genes of his first (resp. second) parent with the exception of r genes that they are randomly selected from the second (resp. first) parent.

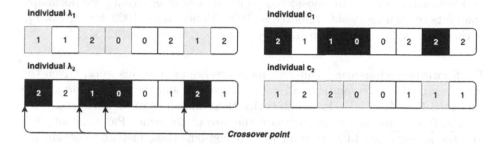

Fig. 4. Example of crossover operator with 8 jobs.

An individual c mutates if a randomly chosen probability is less than or equal to the parameter ρ_{mut}. Two genes (jobs) are randomly selected and their characteristics are swapped (see Fig. 5).

The stopping criterion is a number of iterations.

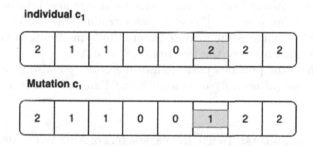

Fig. 5. Example of mutation operator with 8 jobs.

7 Computational Experiments

We implemente our algorithms in Java language and experiments have been driven on a workstation with a 2.2 Ghz Intel Core i7 processor and 16 GB of memory. We use IBM ILOG CPLEX Optimization Studio version 12.0.3 to solve the ILP model.

Data Generation: We assess the performance of the algorithms on 120 instances, with $m \in \{2, 4, 6\}$ and $n \in \{20, 40, 60, 100\}$. 50% of n are jobs of agent A (10 instances are generated per n), for a total of $3 \times 4 \times 10 = 120$

instances. By choosing 50% of jobs for each agent, we are therefore interesting in solving the most difficult problems.

Job-starting dates have been randomly generated according to a discrete uniform distribution between 1 mn and 1439 mn. Similarly, job-finishing dates have been randomly generated between $(s_j + 1)$mn and 1440 mn. Three types of additional resources are considered. Without lost of generality, we normalize the units of each renewable resource to 1000. Hence, $R_k = 1000, k = 1, 2, 3$. r_{jk} is then randomly generated in $[1, 1000]$, $j = 1, \ldots, n$ and $k = 1, 2, 3$.

Performance Measures: Different measures are used in this work. Given the set of optimal Pareto solutions $S^* = \{a_1, \ldots, a_{|S^*|}\}$ and the set of approximate solutions $S = \{b_1, \ldots, b_{|S|}\}$, we categorize these measures in three classes:

Cardinality measures: we calculate the size of the exact Pareto front $|S^*|$, and the approximated Pareto front $|S|$. We combine these metrics to obtain the percentage of strict non-dominated solutions generated by NSGA-II and greedy heuristics: $\%Sol = \frac{|S \cap S^*|}{|S^*|} \times 100$. We also combine these metrics to obtain the percentage of weak non-dominated solutions generated by proposed methods, denoted $\%wSol$.

Average distance: we use the average minimum Euclidian distance: $GD = \frac{1}{|S|}(\sum_{i=1}^{|S|} d_i)$, d_i is the minimum Euclidian distance between the element $b_i \in S$ and the nearest element in S^*.

Hyper-volume: even when the convex hull of the Pareto front S is near to the convex hull of the Pareto-front S^*, the average distance may be a poor indicator of the quality of the front S. Therefore we introduce the metric $HyperV$ for hyper-volume; it calculates the area dominated by some front. In the following, we indicate the area between the two fronts S^* and S.

In addition to the previous performance measures, we also give the average computation time (in second) to compute the Pareto front; it is denoted by CPU.

Fixing the Parameters: To fix the parameters of NSGA-II a subset of twelve instances has been used ($\{20, 40, 60, 100\} \times \{2, 4, 6\}$). The tests show that the best results are obtained when: $N_{\mathcal{P}0} = n + \frac{n}{2}$; The probability of crossover σ_{cross} is set to 0.8; 4 is the number of candidates where one of them will be chosen by tournament selection and will then be one of the two future parents; The probability of mutation ρ_{mut} is set to 0.8; The number of iterations depends on the number of jobs and it is given by $35n$.

Numerical Results: The results are presented in the following table and figures. For each instance, we compare the exact front S^* (computed by the ILP) and the Pareto front S (computed by NSGA-II algorithm and heuristics). Figure 6 indicates the number of Pareto solutions obtained by the exact method, NSGA-II and heuristics with respect number of jobs and number of machines. Hence when the number of jobs and machines increases, the number of Pareto solutions also increases.

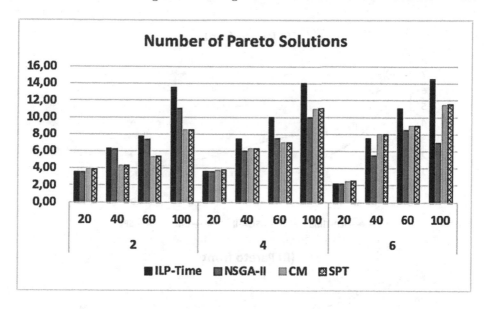

Fig. 6. Average of the number of Pareto solutions obtained by each method.

Note that for some instances the number of non-dominated solutions given by applying heuristics is greater than those given by applying exact method or NSGA-II algorithm. However, some of these points are strictly dominated by the exact solutions (see Fig. 7). On average, the cardinality of the Pareto front is less than 15 solutions. With small size instances (20 jobs), we remark that NSGA-II returns exact Pareto front (see Fig. 7(A)). This performance deteriorates with the increase in the number of jobs. Nevertheless NSGA-II allows to obtain weakly Pareto solutions for each instance (Fig. 7(B) shows this trend). These weakly solutions are very close to the optimal solutions.

From Table 3 we remark that the ILP model needs more than one hour to solve instances with $n = 80$ and $m = 6$. Nevertheless, it remains quite efficient for small instances (less than 1 min for instances with 40 jobs). Dealing with average distance measure (GD), it can be concluded that the distance from the optimal solution set is quit small. For example, the average distance between exact front and approximate front obtained by NSGA-II is less than 2 for instances of 100 jobs and 6 machines. For instances with 40 jobs and 6 machines NSGA-II does not find strict Pareto solutions but all most of returned solutions are weakly Pareto solution (between 96% and 100%). These conclusions are confirmed by the Hyper-volume metric. We can conclude that the approximate Pareto fronts obtained by NSGA-II are close to the exact Pareto fronts. This is not the case when we use greedy heuristics. Note that these types of greedy heuristics are mainly used to decide on the execution of jobs in a highly competitive environment such as a data center or grid computing context.

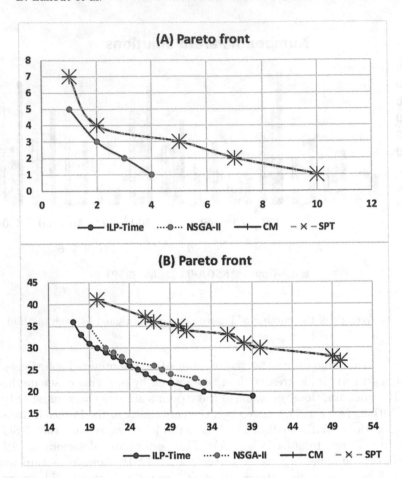

Fig. 7. Example of Pareto front with: (A) $n = 20$ and $m = 4$; (B) $n = 100$ and $m = 4$.

Table 3. Computational results with $n_A = 50\% n$

m	n	ILP		NSGA−II						CM						SPT					
		CPU_s	$|S^*|$	CPU_s	$|S|$	GD	HyperV	%S	%wS	CPU_s	$|S|$	GD	HyperV	%S	%wS	CPU_s	$|S|$	GD	HyperV	%S	%wS
2	20	2,29	3,50	1,50	3,50	0,00	0,00	100,00	0,00	0,04	3,90	1,91	17,50	13,33	66,83	0,01	3,90	1,91	17,50	13,33	66,83
	40	8,22	6,30	5,31	6,20	0,00	1,40	98,57	1,43	0,09	4,30	2,91	63,50	2,50	94,64	0,02	4,30	2,91	63,50	2,50	94,64
	60	15,08	7,67	10,91	7,33	0,30	16,00	71,30	25,00	0,10	5,33	5,36	173,00	0,00	93,33	0,03	5,33	5,36	173,00	0,00	93,33
	100	56,36	13,50	24,97	11,00	0,33	106,50	59,62	40,38	0,20	8,50	7,01	394,00	0,00	100,00	0,06	8,50	7,01	394,00	0,00	100,00
4	20	4,74	3,60	2,37	3,60	0,00	0,00	100,00	0,00	0,06	3,80	1,21	6,60	28,33	64,67	0,01	3,80	1,21	6,60	28,33	64,67
	40	21,38	7,40	8,99	6,00	0,35	5,20	55,26	43,08	0,15	6,30	3,81	87,30	1,25	84,69	0,05	6,30	3,81	87,30	1,25	84,69
	60	52,15	10,00	19,48	7,50	0,43	24,50	55,00	45,00	0,19	7,00	5,95	214,00	0,00	92,86	0,06	7,00	5,95	214,00	0,00	92,86
	100	205,23	14,00	140,04	10,00	1,53	31,00	0,00	100,00	0,33	11,00	8,50	544,50	0,00	72,73	0,16	11,00	8,50	544,50	0,00	72,73
6	20	5,35	2,20	2,80	2,20	0,00	0,00	100,00	0,00	0,06	2,50	0,99	1,50	45,00	42,50	0,01	2,50	0,99	1,50	45,00	42,50
	40	35,55	7,50	11,41	5,50	0,46	4,20	40,00	58,33	0,20	8,00	3,99	83,10	6,51	72,30	0,07	8,00	3,99	83,10	6,51	72,30
	60	95,83	11,00	65,99	8,50	1,12	17,50	4,55	95,45	0,25	9,00	7,14	201,50	0,00	88,89	0,09	9,00	7,14	201,50	0,00	88,89
	100	4510,21	14,50	213,03	7,00	1,98	66,00	0,00	100,00	0,37	11,50	9,57	614,50	0,00	83,46	0,26	11,50	9,57	614,50	0,00	83,46

8 Conclusions

In this paper, we tackled the scheduling problems where two interfering agents compete to perform their jobs on common identical parallel machines. Each agent is associated with a disjoint subset of jobs. Each agent needs to maximize the number of his scheduled jobs. Each objective function depends only on the jobs of the concerned agent. Even when the agents aim at maximizing the same criterion, the problem remains difficult. In our work, we seek to generate the set of Pareto solutions where ε-constraint approach is used for ILP and greedy heuristics. The studied problem admits a polynomial number of Pareto solutions. We then proposed NSGA-II with an adequate encoding scheme and crossover operator. All proposed approaches are compared with time-indexed ILP formulation. The NSGA-II and ILP methods can be generalized to the case of more than two agents and by considering other regular criteria.

One of the applications of our work concerns data centers where the number of jobs can be very large. The NSGA-II offers the best compromise between solution quality and computation time when the data are being controlled. Nevertheless, it will be interesting to develop other methods providing high-quality solution with a low computational running time to solve the online case.

References

1. Agnetis, A., Billaut, J.C., Gawiejnowicz, S., Pacciarelli, D., Soukhal, A.: Multiagent Scheduling: Models and Algorithms. Springer, Heidelberg (2014). https://doi.org/10.1007/978-3-642-41880-8
2. Agnetis, A., Mirchandani, P., Pacciarelli, D., Pacifici, A.: Nondominated schedules for a job-shop with two competing users. Comput. Math. Organ. Theory 6(2), 191–217 (2000). https://doi.org/10.1023/A:1009637419820
3. Agnetis, A., Pacciarelli, D., de Pascale, G.: A Lagrangian approach to single-machine scheduling problems with two competing agents. J. Sched. 12, 401–415 (2009). https://doi.org/10.1007/s10951-008-0098-0
4. Agnetis, A., de Pascale, G., Pranzo, M.: Computing the Nash solution for scheduling bargaining problems. Int. J. Oper. Res. 1, 54–69 (2009)
5. Angelelli, E., Bianchessi, N., Filippi, C.: Optimal interval scheduling with a resource constraint. Comput. Oper. Res. 51, 268–281 (2014)
6. Balasubramanian, H., Fowler, J., Keha, A., Pfund, M.: Scheduling interfering job sets on parallel machines. Eur. J. Oper. Res. 199, 55–67 (2009)
7. Cheng, T.C.E., Ng, C., Yuan, J.J.: Multi-agent scheduling on a single machine to minimize total weighted number of tardy jobs. Theoret. Comput. Sci. 362, 273–281 (2006)
8. Cordeiro, D., Dutot, P.F., Mounié, G., Trystram, D.: Tight analysis of relaxed multi-organization scheduling algorithms. In: Proceedings of the 25th IEEE International Parallel & Distributed Processing Symposium (IPDPS), Anchorage, AL, USA, pp. 1177–1186. IEEE Computer Society (2011)
9. Deb, K., Pratap, A., Agarwal, S., Meyarivan, T.: A fast and elitist multiobjective genetic algorithm: NSGA-II. IEEE Trans. Evol. Comput. 6(2), 182–197 (2002)

10. Elvikis, D., Hamacher, H., T'Kindt, V.: Scheduling two agents on uniform parallel machines with makespan and cost functions. J. Sched. **14**, 471–481 (2011). https://doi.org/10.1007/s10951-010-0201-1
11. Hoogeveen, H.: Multicriteria scheduling. Eur. J. Oper. Res. **167**, 592–623 (2005)
12. Konak, A., Coit, D.W., Smith, A.E.: Multi-objective optimization using genetic algorithms: a tutorial. Reliab. Eng. Syst. Saf. **91**(9), 992–1007 (2006)
13. Kovalyov, M., Oulamara, A., Soukhal, A.: Two-agent scheduling on an unbounded serial batching machine. J. Sched. (2012)
14. Kung, H., Luccio, F., Preparata, F.P.: On finding the maxima of a set of vectors. J. Assoc. Comput. Mach. **22**(4), 469–476 (1975)
15. Peha, J.: Heterogeneous-criteria scheduling: minimizing weighted number of tardy jobs and weighted completion time. Comput. Oper. Res. **22**(10), 1089–1100 (1995)
16. Sadi, F., Soukhal, A.: Complexity analyses for multi-agent scheduling problems with a global agent and equal length jobs. Discrete Optim. **23**, 93–104 (2017)
17. Sadi, F., Soukhal, A., Billaut, J.C.: Solving multi-agent scheduling problems on parallel machines with a global objective function. RAIRO - Oper. Res. **48**(2), 255–269 (2014)
18. T'Kindt, V., Billaut, J.C.: Multicriteria Scheduling. Theory, Models and Algorithms, 2nd edn. Springer, Heidelberg (2006). https://doi.org/10.1007/b106275
19. Tuong, N.H., Soukhal, A., Billaut, J.C.: Single-machine multi-agent scheduling problems with a global objective function. J. Sched. **15**, 311–321 (2012). https://doi.org/10.1007/s10951-011-0252-y
20. Wan, G., Leung, J.Y., Pinedo, M.: Scheduling two agents with controllable processing times. Eur. J. Oper. Res. **205**, 528–539 (2010)
21. Zahout, B., Soukhal, A., Martineau, P.: Fixed jobs scheduling on a single machine with renewable resources. In: MISTA 2017, Kuala Lumpur, Malaysia, pp. 1–9 (2017)

A Study of Recombination Operators for the Cyclic Bandwidth Problem

Jintong Ren[1], Jin-Kao Hao[1,2(✉)], and Eduardo Rodriguez-Tello[3]

[1] LERIA, Université d'Angers, 2 Boulevard Lavoisier, 49095 Angers, France
jin-kao.hao@univ-angers.fr
[2] Institut Universitaire de France, 1 rue Descartes, 75231 Paris, France
[3] Cinvestav - Tamaulipas, Km. 5.5 Carretera Victoria - Soto La Marina,
87130 Victoria Tamps., Mexico

Abstract. This work is dedicated to a study of the NP-hard Cyclic Bandwidth Problem with the paradigm of memetic algorithms. To find out how to choose or design a suitable recombination operator for the problem, we study five classical permutation crossovers within a basic memetic algorithm integrating a simple descent local search procedure. We investigate the correlation between algorithmic performances and population diversity measured by the average population distance and entropy. This work invites more research to improve the two key components of the memetic algorithm: reinforcement of the local search and design of a meaningful recombination operator suitable for the problem.

Keywords: Recombination operators · Memetic algorithms · Cyclic bandwidth · Population diversity

1 Introduction

The Cyclic Bandwidth Problem (CBP) is a typical graph labeling problem, which was introduced in [14] in the context of designing a ring interconnection network. CBP involves finding a disposition of computers on a cycle to make sure that the intercommunication information reaches its destination within at most k steps. The decision version of the problem is known to be a NP-complete problem [15]. In addition to network design, CBP has other relevant applications in very-large scale integration design [3] and data structure representation [25].

CBP can be stated formally as follows: let $G(V, E)$ be a finite undirected graph and C_n a cycle graph, where V $(|V| = n)$ is the set of vertices (or nodes) and $E \subset V \times V$ is the set of edges. Given a bijection (or arrangement) $\varphi : V \to V$ which represents an embedding of G in C_n, the cyclic bandwidth (the cost) of φ for G is defined as,

$$B_C(G, \varphi) = \max_{(u,v) \in E} \{|\varphi(u) - \varphi(v)|_n\}, \tag{1}$$

© Springer Nature Switzerland AG 2020
L. Idoumghar et al. (Eds.): EA 2019, LNCS 12052, pp. 177–191, 2020.
https://doi.org/10.1007/978-3-030-45715-0_14

where $|x|_n = min\{|x|, n - |x|\}$ $(1 \leq |x| \leq n - 1)$ is called the *cyclic distance*, and $\varphi(u)$ denotes the label associated to vertex u. The goal of CBP is to find an arrangement φ^* with minimal $B_C(G, \varphi^*)$.

As a well-known meta-heuristic framework [12,17], memetic algorithms (MAs) have been widely used to solve a large number of NP-hard problems [5,11,13,28,29]. For permutation problems, MAs have also reported good performances for the Traveling Salesman Problem (TSP) [8,16], the Quadratic Assignment Problem [2], and other bandwidth problems [1,20].

Despite the theoretical and practical relevance of CBP, few studies can be found in the literature for solving the problem. A branch and bound algorithm was presented [24] to handle small graphs $(n < 40)$. A tabu search algorithm was proposed [23] to deal with standard and random graphs with 8 to 8192 vertices. Very recently, an iterated three-phase search approach [19] was introduced and improved a number of previous best results reported in [23]. To our knowledge, the memetic approach has never been experimented to solve CBP in the literature, though MAs have been applied to other labeling problems such as the cyclic bandwidth sum problem [22] and the antibandwidth problem [20]. This work fills the gap by investigating the memetic approach for CBP. In particular, we focus on the role of the recombination or crossover (used interchangeably in this paper) component and study the contributions of five permutation recombination operators which are conveniently applicable to CBP. To highlight the impacts of the studied recombination operators, we base our study on a canonical memetic algorithm which combines a recombination operator for solution generation and a simple descent local search for solution improvement.

2   Memetic Algorithm for CBP

2.1   Search Space, Representation, Fitness Function

Given a graph $G = (V, E)$ of order $|V| = n$ and a cycle graph C_n, the search space Ω for the CBP is composed of all possible embeddings (labellings, solutions or arrangements) of G in C_n, $\varphi : V \rightarrow V$. Considering the symmetry characteristic of solutions, there exist $(n - 1)!/2$ possible embeddings for G [23].

Figure 1 shows a graph with 6 vertices named from 'a' to 'f' (Fig. 1(a)). According to Eq. (1), the objective value of Fig. 1(b) is 3 (decided by the longest edge 'dc' in the example). An embedding arranged in a cycle graph (Fig. 1(b)) where the numbers in red are the labels assigned to the vertices, and two embeddings where the vertices are rearranged in the cycle graph in clockwise direction (Fig. 1(c)) and in anticlockwise direction (Fig. 1(d)). Notice that the relative position of each pair of nodes in Fig. 1(b)–(d) is not changed. So according to Eq. (1), these three embeddings have the same objective value, and in fact they correspond to the same solution.

In practice, we represent an embedding φ by permutations $l = \{1, 2, \ldots, n\}$ such that the i-th element $l[i]$ denotes the label assigned to vertex i of V. Another representation of an embedding is proposed in [21], which maps a permutation φ to an array γ indexed by the labels. The i-th value of $\gamma[i]$ indicates the vertex

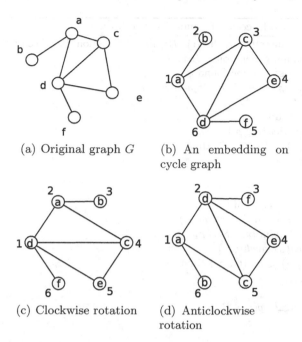

(a) Original graph G

(b) An embedding on cycle graph

(c) Clockwise rotation

(d) Anticlockwise rotation

Fig. 1. Illustration of a graph (a) with an embedding (b) and two equivalent symmetric embeddings (c) and (d)

whose label is i. We illustrate these representations with an example. For the embedding of Fig. 1(b), we have $\varphi = (1\ 2\ 3\ 6\ 4\ 5)$ for the vertices from 'a' to 'f', and the corresponding γ representation is $\gamma = (a\ b\ c\ e\ f\ d)$. In our algorithm, the φ representation is used in the local search procedure, because it eases the implementation of the *swap* operation, while the γ representation is adopted for the recombination operators, as well as the distance calculation presented in Sect. 5. The fitness of a candidate embedding φ in the search space is evaluated by Eq. (1).

2.2 General Procedure

The studied MA follows the general MA framework in discrete optimization [10]. Staring with an initial population (Sect. 2.3), it alternates between a local search procedure (Sect. 2.4) and a recombination procedure (Sect. 2.5). The pseudo-code of the proposed MA is presented in Algorithm 1. The algorithm first fills the population P with $|P|$ local optimal solutions provided by the local search procedure and then performs a series of generations. At each generation, two parent solutions φ_F and φ_M are selected at random from the population and are recombined to generate an offspring solution φ_C. Then, the local search is used to improve the offspring solution to attain a new local optimal solution. Finally, the improved solution is used to update the population (Sect. 2.6). This process is repeated until a fixed number of generations (*MaxGene*) is reached.

Algorithm 1. Pseudo-code of general procedure

1: **Input**: Finite undirected graph $G(V, E)$, fitness function B_C, fixed size of population $|P|$ and maximum generations $MaxGene$
2: **Output**: The best solution found φ^*
3: $P = \{\varphi^1, \varphi^2, ...\varphi^{|P|}\} \leftarrow Init\_Population()$
4: $\varphi^* \leftarrow Best(P)$
5: **for** $i = 1$ to $|P|$ **do**
6:     $\varphi^i \leftarrow Local\_Search(\varphi^i)$
7:     **if** $B_C(G, \varphi^i) < B_C(G, \varphi^*)$ **then**
8:         $\varphi^* \leftarrow \varphi^i$
9:     **end if**
10: **end for**
11: **for** $j = 1$ to $MaxGene$ **do**
12:     $\varphi_F, \varphi_M \leftarrow Parent\_Selection(P)$
13:     $\varphi_C \leftarrow Recombination\_Sol(\varphi_F, \varphi_M)$
14:     $\varphi_C \leftarrow Local\_Search(\varphi_C)$
15:     **if** $B_C(G, \varphi_C) < B_C(G, \varphi^*)$ **then**
16:         $\varphi^* \leftarrow \varphi_C$
17:     **end if**
18:     $P \leftarrow Update\_Pop(\varphi_C, P)$
19:     $j \leftarrow j + 1$
20: **end for**
21: **return** φ^*

2.3   Initialization

In the initialization procedure ($Ini\_Population$), $|P|$ embeddings are generated randomly and independently at first. And then each embedding is improved by the local search procedure of Sect. 2.4 to attain a local optimum (lines 5–10, Algorithm 1). The best solution φ^* in P is also recorded, which is updated during the subsequent search, each time an improved best solution is encountered.

2.4   Local Search

Local search (LS) is an important component of the memetic algorithm, which aims to improve the input solution by searching a given neighborhood. In this work, we apply a simple Descent Local Search (DLS) in order to highlight the contributions of the recombination component.

DLS adopts the swap-based neighborhood of [23], where a neighboring solution of a given solution φ is obtained by simply swapping the labels of two vertices of φ. To specify the neighborhood, we first define, for a vertex u, its cyclic bandwidth $B_C(u, \varphi)$ with respect to the embedding φ as follows:

$$B_C(u, \varphi) = \max_{v \in A(u)} \{|l(u) - l(v)|_n\}, \tag{2}$$

where $A(u)$ denotes the set of vertices adjacent to u of cardinality $deg(u)$. Then the set of critical vertices is given by:

$$C(\varphi) = \{u \in V : B_C(u, \varphi) = B_C(G, \varphi)\}. \qquad (3)$$

The neighborhood is defined as follows:

$$N(\varphi) = \{\varphi' = \varphi \oplus swap(u, v) : u \in C(\varphi), v \in V\}. \qquad (4)$$

DLS starts with an input embedding, then it iteratively visits a series of neighboring solutions of increasing quality according to the given neighborhood. At each iteration, only solutions with a better objective value are considered and the best one is used to replace the incumbent solution. If there exist multiple best solutions, the first one encountered is adopted. We repeat this process until no better solution exists in the neighborhood. In this case, DLS attains a local optimum and the procedure of recombination is triggered to escape from the local optimum.

2.5   Recombination

Recombination is another important part of the MA, which aims to generate new diversified and potentially improving solutions. In our case, only one offspring solution is generated at each generation by each recombination application. In Sect. 3, we present five permutation recombination operators applied to CBP.

2.6   Updating Population

Each new offspring solution improved by the local search procedure is used to update the population. In the proposed MA, we apply a simple strategy: we insert the new offspring into P, and remove the "worst" solution in terms of the objective value.

3   Recombination Operators

There are several recombination operators that are already applied to permutation problems [6,8,9,18,26]. We consider five crossover operators introduced below. It is worth noting that all the recombination operations work with the γ representation mentioned in Sect. 2.1.

3.1   Order Crossover

The Order Crossover operator (OX) [6] generates an offspring solution with a substring of one parent solution and conserves the relative order of the numbers of the other parent solution. Let's consider an example with two parent solutions $\varphi_F = (1\ 2\ 3\ 4\ 5\ 6\ 7\ 8)$ and $\varphi_M = (2\ 4\ 6\ 8\ 7\ 5\ 3\ 1)$ (each number here denotes the index of a node). Given two random cut points (in this case, the first cut point is between second and third positions and the second cut point is between fifth and sixth positions, i.e., $\varphi_F = (1\ 2 \mid 3\ 4\ 5 \mid 6\ 7\ 8)$ and $\varphi_M = (2\ 4 \mid 6\ 8\ 7 \mid 5\ 3\ 1)$, two offspring solutions first inherit the substring between the two cut points: $\varphi_{C1} = (+ + \mid 3\ 4\ 5 \mid + + +)$ and $\varphi_{C2} = (+ + \mid 6\ 8\ 7 \mid + + +)$. Then, we copy the permutation starting from the second cut point of φ_M to the end, as well as from the beginning to the second cut point: $(5\ 3\ 1\ 2\ 4\ 6\ 8\ 7)$. At last, the new obtained permutation is used to insert into φ_{C1} from the second cut point. The repeated numbers are skipped and we get $\varphi_{C1} = (8\ 7 \mid 3\ 4\ 5 \mid 1\ 2\ 6)$. The same operations are performed on φ_{C2} with φ_F to get $\varphi_{C2} = (4\ 5 \mid 6\ 8\ 7 \mid 1\ 2\ 3)$.

3.2   Order-Based Crossover

The Order-based Crossover operator (OX2) [26] is a modified version of OX. Instead of choosing two cut points, OX2 chooses several random positions of one parent solution, and then the order of the selected positions is imposed on the other parent solution. For instance, we have two parent solutions $\varphi_F = (1\ 2\ 3\ 4\ 5\ 6\ 7\ 8)$ and $\varphi_M = (2\ 4\ 6\ 8\ 7\ 5\ 3\ 1)$, and the second, third and sixth positions are picked for φ_M. So the order of "4 6 5" is kept. For solution φ_F, we remove the corresponding numbers of these positions to get $(1\ 2\ 3 + + + 7\ 8)$. Then we insert the numbers in the order "4 6 5" into φ_F and we get the offspring solution $\varphi_{C1} = (1\ 2\ 3\ 4\ 6\ 5\ 7\ 8)$. The same operation can be performed for φ_M to obtain the other offspring solution $\varphi_{C2} = (2\ 4\ 3\ 8\ 7\ 5\ 6\ 1)$.

3.3   Cycle Crossover

The Cycle Crossover operator (CX) [18] seeks a way to preserve the common information in both parent solutions. Two new offspring solutions φ_{C1} and φ_{C2} are created from two parents φ_F and φ_M where the number of each position in φ_{C1} and φ_{C2} is decided by the number of the corresponding position of one parent. For example, we consider two parent solutions $\varphi_F = (1\ 2\ 3\ 4\ 5\ 6\ 7\ 8)$ and $\varphi_M = (2\ 4\ 6\ 8\ 7\ 5\ 3\ 1)$. Firstly, the number of the first position of φ_{C1} could be 1 or 2, Supposing that we pick 1 here $(1 + + + + + + +)$. Then, the number of the eighth position could not be 1 because it is already assigned to the first position, hence we allocate it with a number from φ_F to get $(1 + + + + + + 8)$. After that, we find the position of φ_M whose number is 8 and assign the number of φ_F to the corresponding position of φ_{C1}. We repeat the same operation and find that the forth and the second number of φ_{C1} come from φ_F, which leads to $(1\ 2 + 4 + + + 8)$. For the rest of the positions, we fill them with the numbers from φ_M to obtain a complete offspring solution $\varphi_{C1} = (1\ 2\ 6\ 4\ 7\ 5\ 3\ 8)$. Similarly, we could get the other offspring solution $\varphi_{C2} = (2\ 4\ 3\ 8\ 5\ 6\ 7\ 1)$.

3.4   Partially Mapped Crossover

The Partially Mapped Crossover operator (PMX) [9] passes the absolute position information from the parent solutions to the offspring solutions. An offspring solution gets a substring from one parent and its remaining positions take the values of the other parent. For example, we consider again $\varphi_F = (1\ 2\ 3\ 4\ 5\ 6\ 7\ 8)$ and $\varphi_M = (2\ 4\ 6\ 8\ 7\ 5\ 3\ 1)$. At the beginning, two random cut points are chosen for both parent solutions: $\varphi_F = (1\ 2\ 3\ |\ 4\ 5\ 6\ |\ 7\ 8)$ and $\varphi_M = (2\ 4\ 6\ |\ 8\ 7\ 5\ |\ 3\ 1)$. Then we pass the information between the two cut points to the offspring solutions: $\varphi_{C1} = (+\ +\ +\ |\ 4\ 5\ 6\ |\ +\ +)$ and $\varphi_{C2} = (+\ +\ +\ |\ 8\ 7\ 5\ |\ +\ +)$. Also, we get the mapping for the substrings between the two cut points: $4\leftrightarrow8$, $5\leftrightarrow7$, $6\leftrightarrow5$. After that, the other positions of the offspring solutions are filled with the other parent solution, hence we get $\varphi_{C1} = (2\ 4\ 6\ |\ 4\ 5\ 6\ |\ 3\ 1)$ and $\varphi_{C2} = (1\ 2\ 3\ |\ 8\ 7\ 5\ |\ 7\ 8)$. For the duplicate labels in the solution, we use the mapping of substrings to replace the repeated numbers. In this case, $5\leftrightarrow7$ and $6\leftrightarrow5$ result in $6\leftrightarrow7$. Therefore, the offspring solutions are $\varphi_{C1} = (2\ 8\ 7\ |\ 4\ 5\ 6\ |\ 3\ 1)$ and $\varphi_{C2} = (1\ 2\ 3\ |\ 8\ 7\ 5\ |\ 6\ 4)$.

3.5 Distance Preserved Crossover

The Distance Preserved Crossover operator (DPX) [8], designed for solving the Traveling Salesman Problem (TSP), aims to produce an offspring solution which has the same distance to each of its parents. It is noteworthy that the distance here is the distance based on the common connections between two solutions, instead of the Hamming distance. We come back to this issue in Sect. 5. For DPX, we firstly delete the uncommon connections of two neighboring numbers for both parent solutions. Then, the parent solutions are separated into different substrings. Finally, we reconnect all the substrings without using any of the connections which are contained in only one of the parent solutions. For more detailed explanations and examples, please refer to [8].

4 Experimental Results

4.1 Instances and Settings

In this section, we report experimental results of the MA using the 5 different recombination operators introduced in Sect. 3. The study was based on 20 representative graphs with 59 to 2048 vertices, selected from a test-suite of 113 benchmark instances (https://www.tamps.cinvestav.mx/~ertello/cbmp.php). 14 of the chosen graphs are standard graphs covering 7 dissimilar categories (path, cycle, complete tree, 2-dimension mesh, 3-dimension mesh, caterpillar and hypercube) and the other 6 graphs (called Harwell-Boeing graphs) come from real-world scientific and engineering applications and are part of the Harwell-Boeing Sparse Matrix Collection. Considering the stochastic nature of the algorithm, each instance was independently solved 50 times under the environment of Linux using an Intel Xeon E5-2695 2.1 GHz CPU and 2 GB RAM. Each execution was limited to 20000 generations ($MaxGene = 20000$) and the population size $|P|$ was set to 20.

4.2   Computional Results

Table 1 outlines the computational results of our MA variants with the 5 different recombination operators. The columns "Best" and "Avg" list the best and average objective values found. According to the definition introduced in Sect. 1, a smaller objective value indicates a better result. Table 1 shows that the algorithm with OX2 obtains the best results not only in terms of "Best" but also in terms of "Avg" over the 20 test instances. From the average values listed in the last row, we find that OX2 is a much more suitable operator than the other operators for CBP. Also, the non-parametric Friedman test on the 5 groups of best results leads to a p-value $= 6.71e\text{-}14 < 0.05$, confirming that there exists a statistically significant difference among the compared results.

Table 1. Experimental results of MA using 5 different recombination operators.

Graph	CX		DPX		OX		OX2		PMX	
	Best	Avg	Best	Avg	Best	Avg	Best	Avg	Best	Avg
nos6	327	331.28	327	329.74	266	287.98	**216**	**227.84**	327	331.98
path1000	461	475.42	462	474.02	254	301.04	**226**	**247.54**	468	482.68
nos4	44	46.12	43	45.24	32	39.32	**28**	**34.48**	42	45.78
tree10x2	39	42.72	35	40.72	**28**	32.50	**28**	**29.26**	36	41.56
cycle1000	457	476.66	466	473.38	252	296.98	**226**	**246.94**	459	480.86
mesh2D8x25	88	93.04	89	91.82	59	75.18	**57**	**62.94**	87	93.28
caterpillar29	203	211.48	203	208.70	138	162.98	**100**	**127.32**	198	210.14
mesh3D6	102	103.88	101	102.96	86	93.08	**73**	**78.26**	102	104.28
hypercube11	1022	1022.76	1022	1022.14	1019	1021.26	**952**	**1010.48**	1022	1022.54
cycle475	200	215.16	206	213.36	105	128.36	**99**	**110.76**	192	217.30
mesh2D28x30	409	413.40	410	412.06	336	371.76	**270**	**287.46**	406	414.06
mesh3D11	660	662.04	660	661.28	625	650.30	**507**	**522.82**	660	662.40
can_715	354	355.80	355	355.14	347	353.92	**293**	**316.70**	354	355.74
impcol_b	28	28.46	27	27.96	25	27.22	**20**	**26.72**	28	28.00
path475	202	214.50	206	212.86	112	132.24	**102**	**112.94**	189	217.56
494_bus	220	230.76	222	228.72	135	165.74	**128**	**138.62**	216	233.38
tree21x2	199	212.08	203	208.96	139	171.34	**124**	**140.84**	200	210.68
caterpillar44	481	493.28	479	491.24	340	400.78	**281**	**321.70**	480	495.60
impcol_d	207	209.60	207	208.80	190	202.98	**159**	**169.74**	208	209.80
tree2x9	475	489.08	478	485.86	296	330.14	**257**	**276.60**	472	491.84
Average	308.90	316.38	310.50	315.75	239.20	262.26	**207.30**	**224.50**	307.30	317.47
p-value	6.71e-14									

Table 2 reports the comparative results between the best MA with OX2 (called MA_{OX2}) and TS_{CB}, which is the state-of-art algorithm for CBP presented in [23]. Table 2 shows the same information as in Table 1, except for the column "CC" which represents the difference between the best values found by TS_{CB} and MA_{OX2}. A negative "CC" indicates a worse result of MA_{OX2} compared to TS_{CB}. It is clear that for the 20 test graphs, MA_{OX2} does not compete well with TS_{CB}. Indeed, TS_{CB} is a powerful iterated tabu search algorithm which uses three dedicated neighborhoods to effectively explore the search space. Also, the Wilcoxon signed-rank test with the two groups of best values leads to a p-value $= 1.31e\text{-}4 < 0.05$, confirming the statistical significance between the compared results. This comparison tends to indicate that in practice, it is not enough for the MA to apply a recombination operator and a simple local search. In addition to a suitable recombination operator, the MA needs a powerful local optimization procedure to ensure an effective exploitation.

Table 2. Comparison between MA_{OX2} and TS_{CB} [23].

| Graph | MA_{OX2} | | TS_{CB} | | |
	Best	Avg	Best	Avg	CC
nos6	216	227.84	22	23.50	−194
path1000	226	247.54	8	8.90	−218
nos4	28	34.48	10	10.00	−18
tree10x2	28	29.26	28	28.00	0
cycle1000	226	246.94	8	8.50	−218
mesh2D8x25	57	62.94	8	8.20	−49
caterpillar29	100	127.32	24	25.80	−76
mesh3D6	73	78.26	31	31.00	−42
hypercube11	952	1010.48	570	582.20	−382
cycle475	99	110.76	5	5.80	−94
mesh2D28x30	270	287.46	30	174.00	−240
mesh3D11	507	522.82	336	336.80	−171
can__715	293	316.70	60	65.80	−233
impcol_b	20	26.72	17	17.00	−3
path475	102	112.94	5	5.60	−97
494_bus	128	138.62	46	56.10	−82
tree21x2	124	140.84	116	116.00	−8
caterpillar44	281	321.70	39	54.00	−242
impcol_d	159	169.74	38	43.10	−121
tree2x9	257	276.60	63	64.20	−194
Average	207.30	224.50	**73.20**	**83.23**	
p-value	1.31e-4				

5   Understanding the Performance Differences of the Compared Crossovers

In Sect. 4, we observe that OX2 excels compared to the other crossover operators. In this section, we investigate the reasons why OX2 has a better performance than the other crossovers. For this, we follow [27] and study the evolution of the population diversity. To this end, we consider two diversity indicators: average solution distance $D_{avg}(P)$ and population entropy $E_p(P)$.

5.1   Distance and Population Entropy

We first introduce the average solution distance $D_{avg}(P)$ of the population.

$$D_{avg}(P) = \frac{2}{|P|(|P| - 1)} \sum_{i=1}^{|P|} \sum_{j=i+1}^{|P|} d_{ij} \tag{5}$$

where d_{ij} is the distance between two solutions γ_i and γ_j of P, which is defined as the number of the adjacent connections that are contained in γ_i but not in γ_j. For example, given two solutions $\gamma_1 = \{$h a b d e f c g$\}$ and $\gamma_2 = \{$b a c h g d f e$\}$. The set of adjacent connections is $\{$ha, ab, bd, de, ef, fc, cg, gh$\}$ for γ_1 and $\{$ba, ac, ch, hg, gd, df, fe, eb$\}$ for γ_2. The common adjacent connections are $\{$ab, ef, gh$\}$ (ba and ab are the same connections). The distance d_{12} equals thus $8-3 = 5$. This distance is used in [8] to deal with TSP whose solutions have the symmetry feature. As shown in Fig. 1, CBP solutions have the feature of symmetry, so the use of this distance measure is very important for CBP.

Another indicator to describe the population diversity is the population entropy $E_p(P)$ [7].

$$E_p(P) = \frac{-\sum_{i=1}^{n} \sum_{j=1}^{n} \left(\frac{n_{ij}}{|P|}\right) \log \left(\frac{n_{ij}}{|P|}\right)}{n \log n} \tag{6}$$

where n_{ij} represents the number of times that variable i is set to value j in all solutions in P. One notices that $E_p(P)$ varies in the interval [0,1]. When $E_p(P)$ equals 0, all the solutions of P are identical. A large $E_p(P)$ value indicates a more diverse population.

The instance 'nos6' is a representative large graph with 675 nodes from practical application and rather difficult, so we use it as a working example. Figure 2 shows the average distance, average entropy and average best objective value found in 50 independent executions over the graph 'nos6'. Under 5000 generations, the population of the MA with OX2 has a high average distance and entropy, leading to much better solutions. From generations 5000 to 20000, the entropy is identical to that of OX, and the best average objective found stops decreasing. These observations remain valid for all test graphs except the graph 'impcol_b' (even if the MA with OX2 does not have a large population distance and entropy, it gets good results comparing to others). Therefore, for the operators CX, OX, OX2 and PMX, a higher entropy and average distance of population leads to a good quality solution. However, what is surprising is that the average distance and entropy with DPX always stay at a high level for all test graphs, yet the quality of solutions found is not as good as that of the other operators. To shed light on this behavior, we show a deeper analysis of the interaction between the crossover mechanism and the characteristics of problem in the next section.

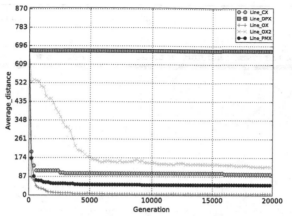

(a) Average distance of the population in 20000 generations

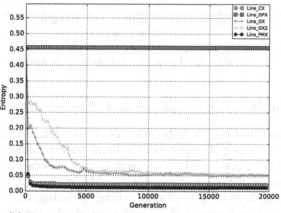

(b) Entropy of the population in 20000 generations

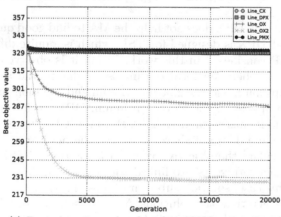

(c) Best objective value found in 20000 generations

Fig. 2. Distance and population entropy applied to the instance nos6.

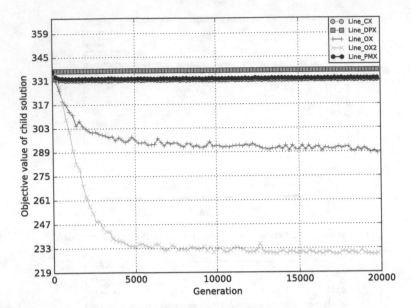

Fig. 3. Average objective value of the child solution over 50 independent executions.

5.2   Interaction Between Crossover and Problem Characteristics

In Sect. 5.1, we find that the recombination operator with a higher entropy and average distance of the population generally helps to find solutions of good quality. However, the DPX operator fails to reach good solutions even if the entropy and average distance of population under the MA with DPX always stay at a high level. From Fig. 3, which presents the average objective value of the offspring solutions of instance nos6 using the average data of 50 independent executions, we find that DPX does not generate high quality offspring solutions during the search.

To understand why DPX does not help the MA to find good quality solutions, we first recall that DPX is designed for TSP, which is a quite different problem compared to CBP considered in this work. In [4], it is observed that for TSP, the average distance between local optima is similar to the average distance between a local optimum and the global optimum and common substrings in the local optima also appear in the global optimum. DPX explores this particular feature of TSP and is thus suitable to TSP. However, CBP has a totally different objective function and does not share the above characteristic.

Indeed, to calculate the objective value of a solution of TSP, we only need to consider, for each vertex, its two linked edges and sum up the edge distances of the tour. In this case, solution sub-tours (substrings) are clearly a key component which characterizes the solutions. Yet in a solution of CBP, we need to consider for each vertex all the edges linked to the vertex in the graph, such that the objective value (see Eq. (1)) relies on the largest cyclic bandwidth. In the case of CBP, the key point is the relative position for the pairs of nodes which are linked by an edge. Therefore given that TSP and CBP have very different characteristics, a good crossover operator designed for TSP (in our case, DPX) may fail when it is applied to CBP.

This inspires us that the choice and design of recombination operators are not only relied on the entropy and average distance of population, but also on the characteristics of the considered problem.

6   Conclusion

In this paper, we have investigated the memetic framework for solving the NP-hard Cyclic Bandwidth problem. We have compared five permutation recombination operators (CX, OX, OX2, PMX and DPX) within a basic memetic algorithm which uses a simple descent procedure for local optimization. The experimental results indicate that OX2 achieves the best performance for the test instances. We have studied the population diversity measured by the average distance and entropy of the MA variants using different recombination operators. We have also explored the correlation between the population diversity and the performance of the studies MA variants. This study indicates that the basic memetic algorithm combining an existing recombination operator and a simple descent local search procedure is not competitive compared to the state-of-the-art CBP algorithms. Additional (preliminary) experiments with MAs using an enforced local optimization procedure (such as the powerful local search algorithms presented in [19,23]) have not led to more convincing results. Meanwhile, given the excellent performances achieved by MAs on many difficult optimization problems, this work invites more research effort on seeking meaningful recombination operators suitable for CBP. It is then expected that a MA integrating such a recombination operator and a powerful local optimization procedure would achieve state-of-the-art performances.

Acknowledgments. We are grateful to the referees for their valuable suggestions and comments which helped us to improve the paper. Support from the China Scholarship Council (CSC, Grant 201608070103) for the first author and support from the Mexican Secretariat of Public Education through SEP-Cinvestav (2019–2020, Grant 00114) for the third author are also acknowledged.

References

1. Bansal, R., Srivastava, K.: A memetic algorithm for the cyclic antibandwidth maximization problem. Soft Comput. **15**(2), 397–412 (2011)
2. Benlic, U., Hao, J.K.: Memetic search for the quadratic assignment problem. Expert Syst. Appl. **42**(1), 584–595 (2015)
3. Bhatt, S.N., Leighton, F.T.: A framework for solving VLSI graph layout problems. J. Comput. Syst. Sci. **28**(2), 300–343 (1984)
4. Boese, K.D.: Cost versus distance in the traveling salesman problem. UCLA Computer Science Department Los Angeles (1995)

5. Chen, Y., Hao, J.K.: Memetic search for the generalized quadratic multiple knapsack problem. IEEE Trans. Evol. Comput. **20**(6), 908–923 (2016)
6. Davis, L.: Applying adaptive algorithms to epistatic domains. In: International Joint Conference on Artificial Intelligence, vol. 85, pp. 162–164 (1985)
7. Fleurent, C., Ferland, J.: Object-oriented implementation of heuristic search methods for graph coloring. Cliques, Coloring, and Satisfiability. DIMACS Ser. Discrete Math. Theor. Comput. Sci. **6**, 619–652 (1996)
8. Freisleben, B., Merz, P.: A genetic local search algorithm for solving symmetric and asymmetric traveling salesman problems. In: Proceedings of IEEE International Conference on Evolutionary Computation, pp. 616–621. IEEE (1996)
9. Goldberg, D.E., Lingle, R., et al.: Alleles, loci, and the traveling salesman problem. In: Proceedings of International Conference on Genetic Algorithms and Their Applications, vol. 154, pp. 154–159. Lawrence Erlbaum, Hillsdale (1985)
10. Hao, J.K.: Memetic algorithms in discrete optimization. In: Neri, F., Cotta, C., Moscato, P. (eds.) Handbook of Memetic Algorithms. Studies in Computational Intelligence, vol. 379, pp. 73–94. Springer, Heidelberg (2012). https://doi.org/10.1007/978-3-642-23247-3_6
11. Jin, Y., Hao, J.K., Hamiez, J.P.: A memetic algorithm for the minimum sum coloring problem. Comput. Oper. Res. **43**, 318–327 (2014)
12. Krasnogor, N., Smith, J.: A tutorial for competent memetic algorithms: model, taxonomy, and design issues. IEEE Trans. Evol. Comput. **9**(5), 474–488 (2005)
13. Lai, X., Hao, J.K.: A tabu search based memetic algorithm for the max-mean dispersion problem. Comput. Oper. Res. **72**, 118–127 (2016)
14. Leung, J.Y., Vornberger, O., Witthoff, J.D.: On some variants of the bandwidth minimization problem. SIAM J. Comput. **13**(3), 650–667 (1984)
15. Lin, Y.: The cyclic bandwidth problem. In: Chinese Science Abstracts Series A, vol. 14(2 Part A), p. 14 (1995)
16. Merz, P., Freisleben, B.: Memetic algorithms for the traveling salesman problem. Complex Syst. **13**, 297–345 (1997)
17. Moscato, P., Cotta, C.: A gentle introduction to memetic algorithms. In: Glover, F., Kochenberger, G.A. (eds.) Handbook of Metaheuristics. International Series in Operations Research & Management Science, vol. 57, pp. 105–144. Springer, Boston (2003). https://doi.org/10.1007/0-306-48056-5_5
18. Oliver, I., Smith, D., Holland, J.: A study of permutation crossover operators on the travelling salesman problem. In: Proceedings of the Second International Conference on Genetic Algorithms and their Application, pp. 224–230 (1987)
19. Ren, J., Hao, J.K., Rodriguez-Tello, E.: An iterated three-phase search approach for solving the cyclic bandwidth problem. IEEE Access **7**, 98436–98452 (2019)
20. Rodriguez-Tello, E., Betancourt, L.C.: An improved memetic algorithm for the antibandwidth problem. In: Hao, J.-K., Legrand, P., Collet, P., Monmarché, N., Lutton, E., Schoenauer, M. (eds.) EA 2011. LNCS, vol. 7401, pp. 121–132. Springer, Heidelberg (2012). https://doi.org/10.1007/978-3-642-35533-2_11
21. Rodriguez-Tello, E., Hao, J.K., Torres-Jimenez, J.: An improved simulated annealing algorithm for bandwidth minimization. Eur. J. Oper. Res. **185**(3), 1319–1335 (2008)
22. Rodriguez-Tello, E., Narvaez-Teran, V., Lardeux, F.: Comparative study of different memetic algorithm configurations for the cyclic bandwidth sum problem. In: Auger, A., Fonseca, C.M., Lourenço, N., Machado, P., Paquete, L., Whitley, D. (eds.) PPSN 2018, Part I. LNCS, vol. 11101, pp. 82–94. Springer, Cham (2018). https://doi.org/10.1007/978-3-319-99253-2_7

23. Rodriguez-Tello, E., Romero-Monsivais, H., Ramirez-Torres, G., Lardeux, F.: Tabu search for the cyclic bandwidth problem. Comput. Oper. Res. **57**, 17–32 (2015)
24. Romero-Monsivais, H., Rodriguez-Tello, E., Ramírez, G.: A new branch and bound algorithm for the cyclic bandwidth problem. In: Batyrshin, I., Mendoza, M.G. (eds.) MICAI 2012, Part II. LNCS (LNAI), vol. 7630, pp. 139–150. Springer, Heidelberg (2013). https://doi.org/10.1007/978-3-642-37798-3_13
25. Rosenberg, A.L., Snyder, L.: Bounds on the costs of data encodings. Math. Syst. Theory **12**(1), 9–39 (1978)
26. Syswerda, G.: Scheduling optimization using genetic algorithms. In: Handbook of Genetic Algorithms, pp. 322–349 (1991)
27. Wang, Y., Lü, Z., Hao, J.-K.: A study of multi-parent crossover operators in a memetic algorithm. In: Schaefer, R., Cotta, C., Kołodziej, J., Rudolph, G. (eds.) PPSN 2010, Part I. LNCS, vol. 6238, pp. 556–565. Springer, Heidelberg (2010). https://doi.org/10.1007/978-3-642-15844-5_56
28. Wu, Q., Hao, J.K.: Memetic search for the max-bisection problem. Comput. Oper. Res. **40**(1), 166–179 (2013)
29. Zhou, Y., Hao, J., Glover, F.: Memetic search for identifying critical nodes in sparse graphs. IEEE Trans. Cybern. **49**(10), 3699–3712 (2019)

Automatic Calibration of a Farm Irrigation Model: A Multi-Modal Optimization Approach

Amaury Dubois[1,2]([✉]), Fabien Teytaud[1]([✉]), Eric Ramat[1],
and Sébastien Verel[1]([✉])

[1] LISIC, Université du Littoral Côte d'Opale,
50 rue Ferdinand Buisson, 62228 Calais, France
dubois.amaury62@gmail.com, fabien.teytaud@gmail.com,
verel@univ-littoral.fr
[2] Weenat Technocampus Alimentation,
2 impasse Therese Bertrand-Fontaine, 44320 Nantes, France
http://www-lisic.univ-littoral.fr/,
https://www.weenat.com/

Abstract. In agriculture, plant cultivation requires to take numerous decisions. One of the major problems is irrigation: an adequate irrigation decision must be made accordingly to the hydric status of the plant and soil, and the weather forecasts. In precision agronomy, this leads to the use of hydric sensors combined with a numerical growth plant model. Such models can not often be tuned by experts. We proposed an automatic parameter calibration of the potato growth model based on data collected in several open fields. As these parameter calibration problems are ill-posed, the associated black-box optimization problem is supposed to be multi-modal. We then compare the performances of two state-of-the-art Evolution Strategies which use different restart mechanisms to automatically tune the set of parameters on different crops and shows that multi-modal optimization methods may be recommended for such class of optimization problems.

Keywords: Multi-Modal Optimization · Real world application · Data driven calibration

1 Introduction

As others domains (industry, urban, etc.), precision agronomy benefits new sensors which are enhanced by numerical models and simulations. Therefore, the decision-making process can be supported by the knowledge bringing by this new numerical environment. For plant cultivation, one major decision is irrigation. The farmer has to decide to irrigate a field according to the hydric state of plant

Supported by WEENAT.

and soil, the weather prediction, and the cost of irrigation. In that case, new decision-making method uses hydric sensors to measure the quantity of humidity of the soil, and, a growth plant model to able to estimate the hydric state of the plant, and the available quantity of water for the plant which depends on the root size, and the characteristics of the soil. Although such numerical approach can lead to an accurate prediction of the crop state, and beyond the sensor precision, one drawback is the setting of the numerous parameters of the plant growth model. Indeed, models combine different sub-models based on differential equations, finite state transitions, etc. that require the settings of many numerical biological, or geological parameters. Even if experts can measure, estimate, or give bounds of some parameter values[1], most of times precise value of parameters are not known for a field-scale crop as they depend on specific soil, and plant species/varieties. In this work, we show that it is possible to set precisely the model parameters of potato plant growth based on the data acquisition of hydric sensors, and a relevant optimization algorithm that minimizes the distance between predicted values computed by the model, and real data.

In evolutionary computation, this black-box problem is known as a calibration problem [1,2]. The parameter settings of the potato plant growth model show specific difficulties. As the parameters depends on local specificity such as soil, potato variety, weather exposition, etc. the data are difficult to collect, and rare: a campaign of data acquisition with hydric sensors for a potato field lasts 4 months, and can be done on the same field every 3 years due to crop rotation. On the other hand, the number of model parameters is high: several dozens for representative models such as STICS [3], AquaCrop [4], or Weedric [5]. As a consequence, the calibration problem of potato growth model is ill-posed. Several parameters settings lead to the same input-output behavior, and the simulations are consistent with the measurements in the field. Thus, the optimization problem related to model calibration is not only a problem with many local optima but a *multi-modal* problem for which the quality of several local optima is very close to those of the global optima.

In this work, we formulate the calibration optimization problem from the farming irrigation model *Weedric* (defined in Sect. 2.1), and available hydric data. More precisely, we have a model with many continuous parameters for which we have no a priori knowledge about their implications in equations, simulations or the interactions between the different model parts. Experts can only define the bounds of each parameter values. So, we consider it as a continuous black-box function from the search space of dimension $d > 1$: $[0, 1]^d$. As stated above, we assume that this problem is highly multi-modal, with one global optimum but has many local optima close to the global one. It is, therefore, necessary to check the maximum number of optima in order to determine the global one.

This kind of problems is known as Multi-Modal Optimization (MMO). Numerous algorithms have been proposed, many of them use the derivative of the gradient but in black box context, these algorithms are not directly

---

[1] Some parameters can also have no meaning from a biological point of view.

applicable. Gradient free methods are generally based on Evolution Strategy (ES) which have shown their robustness and their efficiency [6–8]. ES consists in generating better solutions iteratively from a starting point. In MMO context, ES are generally combined with either a niching technique [9,10] or a restart strategy [11–13], in order to find all the optima. In this paper, we propose to compare different state-of-the-art restart strategy algorithms: *QRDS* [14,15] and *CMAES-IPOP* [12] to automatically tune the parameters model.

The rest of this paper is organized as follows: Sect. 2.1 describes the *Weedric* simulator. Then, we present the Quasi-Random Restart Strategy (QRDS) and Covariance Matrix Adaptation Evolution Strategy with Increasing POPulation (CMAES-IPOP) in Sect. 3. Next, we compare their performances in Sect. 4. Finally we conclude in Sect. 5.

2   Calibration Problem of an Irrigation Model

2.1   Farming Irrigation Model

Many models have been proposed to deal with this plant growth. For example, STICS [3] is a deterministic generic model for the simulation of crops and their water and nitrogen balance developed at INRA institute (France) since 1996. It calculates both agricultural variables (yield, input consumption) and environmental variables (water and nitrogen losses). AquaCrop [4] is also a deterministic generic model. It provides an estimation of crop productivity in relation to water supply and agronomic management in a framework based on current plant physiological and soil water budgeting concepts. Unfortunately, these models require a large number of parameters such as they are generic, several types of plants can be modeled, and involve other biological mechanisms in addition to the irrigation issue.

Weedric is an agricultural irrigation model for the culture of potatoes developed by the *Weenat* company[2]. This model has emerged from a collaborative project between computer, and agriculture researchers [5]. It is intended to farmers in order to help them with decision support.

The Weedric model consists of the combination of several deterministic existing biological models [16–19] to provide a specific model for this kind of culture. Initially, these models are independent, and the interest of the Weedric model (see Fig. 1) is to be able to connect them in order to propose two high-level models:

– Soil model: this model considers the soil as a succession of horizontal layers and each layer has a quantity of water varying over time according to the different exchanges between the layers (percolation, upwelling), the weather (temperature, rain, wind,...) and the interactions with the Plant model.
– Plant model: it simulates the development and the behavior of a potato plant from planting to harvesting, based on current water quantity and weather forecasts.

---

[2] https://www.weenat.com/.

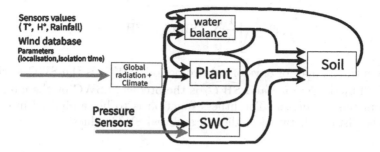

Fig. 1. Simplified diagram of how the multi-model *Weedric* growth crop model works. This one is divided into several sub-models in which the $d = 38$ variables are assigned. The green arrows indicate the entry points of the model by which the different inputs (pressure, temperature, wind, rainfall, ...) can be filled in. The black arrows indicate the interaction and information sharing of the models between them. SWC is the Soil Water Content under interest in this work. (Color figure online)

Using the planting date and the weather forecasts, the multi-model Weedric can predict the water stress of the potato plant, and the Soil Water Content (SWC). The SWC is the available quantity of water that the potato plant can use, and extract from the soil. Unfortunately, to be fully effective, the $d = 38$ real parameter values must be tune for a particular potato variety and soil type.

2.2   Black-Box Calibration Problem

The goal is to calibrate the Weedric model using the Soil Water Content (SWC), and the hydric sensors. Sensors are put in the field, and they regularly send data. This makes it possible to obtain an approximation of the pressure of the water in the ground which can be converted into a quantity of water using the well know Van Gernuchen equation [20]. The black curve "sensors" of the Fig. 2 shows the SWC over a season of $n = 73$ days.

Following the expert knowledge, a set of default parameters value is defined. The green curve "default" on the Fig. 2 shows the predicted SWC by the Weedric using those default values. From the first day to approximately the 30th day, the predicted value follows the measured values by sensors. During this period, the SWC increasing is mainly due to the increasing of the roots. During a dry period after the 30th day, the model with default parameter values seems to over-estimate the dryness of the soil which could be due to misleading values of the soil model, or plant model, or a combination of the two. The interaction of different components of the model are not linear.

The fitness function of the calibration problem is defined as the root mean square error over the crop period of the SWC value predicted by the model. More formally, for every settings $x \in [0,1]^d$, with $d = 38$, of the Weedric model, the fitness function f is defined by:

$$f(x) = \sqrt{\frac{1}{n} \sum_{t=1}^{n} (SWC_t - \widehat{SWC_t})^2} \qquad (1)$$

where n is the number of days of culture period, SWC_t is the SWC at the day t measured by the sensors, and $\widehat{SWC_t}$ is the predicted SWC by the model with the parameters settings x. The fitness function is to be minimized in order to reduce the distance between predicted, and real SWC values.

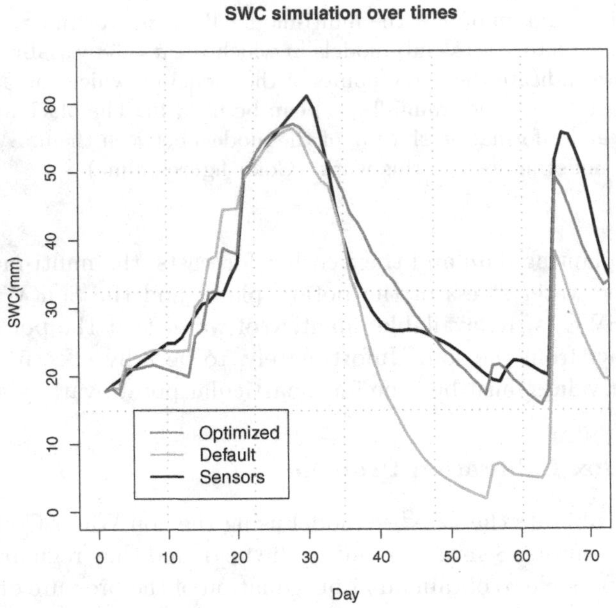

SWC simulation over times

Fig. 2. Evolution of the Soil Water Quantity (SWC) over time ($n = 73$ days). The black curve represents the sensor values, the green represents the prediction of the model with the basic values and the red curve, the prediction of the model with the optimized values. The closer the curves are to the black curve, the better the prediction. At the beginning of the simulation, the two parameter sets are quite similar but over time, the model with the default parameters is no longer accurate. (Color figure online)

3   Multi-Modal Optimization Algorithms

In this section, we present the Quasi-Random restarts with Decreasing Step-size algorithm (QRDS) and the Covariance Matrix Adaptation Evolution Strategy with Increasing POPulation (CMA-ES IPOP), one of its variants for multi-modal problems.

3.1   Quasi-Random Restarts with Decreasing Step-Size

Random restarts with Decreasing Step-size and its improvement (Quasi-Random restarts with Decreasing Step-size [15]) are an Evolution-Strategy-based Multi-Modal Optimization algorithms which use the restarting technique.

It is composed by a simple local search algorithm combined with a restart strategy following a random (or quasi-random) sequence. Local search is a simple (1+1)-ES using the 1/5 adaptation rule (see Algorithm 1).

At the beginning, a point is selected. Then, Iteratively, the algorithm generates a candidate by mutating the current best point according to a normal distribution with a standard deviation (step-size) σ. The best of both points are kept. The update of the step-size σ is really simple: if the candidate solution, i.e. the newly generated point is better, the step-size σ is increased, otherwise, σ is decreased as we may need to focus on smaller neighborhood.

We use this step-size value as the stopping criterion of the local search. If it is too small, we consider that the local search has converged to an optimum (global or local). The solution is saved and the local search is restarted until the evaluation budget is reached.

A feature of QRDS is its *"murder operator"*. In order to avoid converging to an already known solution, the algorithm checks at each evaluation, if the current solution is greater than a distance $\delta_{threshold}$ of all the optima already discovered. If it is true, the search is aborted without saving the solution (we don't want to spend time for an already found optimum).

For the restart strategy (see Algorithm 2), each time the algorithm is restarted, the initial position is sampled according to a quasi-random sequence.

3.2   CMA-ES IPOP

The Covariance Matrix Adaptation Evolution-Strategy is an Evolution Strategy that adapts the full covariance matrix of a normal search distribution [21]. This algorithm is presented in Algorithm 3. An important property of this algorithm is its invariance against linear transformations of the research space. CMA-ES is effective in minimizing unimodal function and is superior when the problem is ill-conditioned and non-separable. In multi-modal context, [22] shows that increasing the size of the population can improve performances of the CMA-ES. [12] proposes a version of CMA-ES using a restart strategy: at each restart (whenever the stopping criterion is met), the size of the population is doubled see Algorithm 4. By increasing the population size, the local search becomes more global after each restart. The results given in [12] show that this improvement provides good performances on multi-modal black-box context.

Algorithm 1: SearchDS

{Search an optimum using a Decreasing Step-size }
Input:
  f: function to optimize
  σ_0: initial step-size
  ϵ_σ: threshold value of the step-size
  y^*: maximum fitness of the function
  ϵ_y: threshold value of the fitness
  $\mathbf{x}$: initial position for the search
  ϵ_x: threshold value of the position
  $\hat{\mathbf{X}}$: set of previously found optima
Output:
  $\hat{\mathbf{X}}$: updated set of optima

```
1  begin
2  |  y ← f(x)
3  |  σ ← σ₀
4  |  repeat
      |     {mutation}
5  |     x' ← N(x, σ)
6  |     y' ← f(x')
      |     {selection with 1/5th adaptation}
7  |     if y' > y then
8  |        | x ← x'
9  |        | σ ← 2σ
10 |     else
11 |        | σ ← 2^(-1/4) σ
      |     {discard search if optimum already known}
12 |     if ∃x̂ ∈ X̂, ||x − x̂|| < εₓ then
13 |        | break
      |     {store found optimum}
14 |     if ||y − y*|| < ε_y then
15 |        | X̂ ← X̂ ∪ {x}
16 |        | break
17 |  until σ < ε_σ or max nb of function evaluations reached
```

4   Experimental Analysis

We compare the multi-modal algorithms, on the *Weedric* model calibration problem defined in the previous Sect. 2. The data from 5 different crops are used with different soil types and potato varieties. The two algorithms are also compared with default parameters values given by the experts. For each crop, the number of independent runs of each algorithm is 100. The maximum number of evaluations for each algorithm is 10^5. Notice that the simulation time of Weedric model is enough short for such number of evaluation within minutes.

Algorithm 2: RDS

{Random restarts with Decreasing Step-size}
Input: $f, \sigma_0, \epsilon_\sigma, y^*, \epsilon_y, \epsilon_x$
Output: $\hat{\mathbf{X}}$

1 **begin**
2 $\quad\hat{\mathbf{X}} \leftarrow \varnothing$
3 $\quad$**repeat**
4 $\quad\quad\mathbf{x} \leftarrow$ sample search-space
5 $\quad\quad\hat{\mathbf{X}} \leftarrow \text{SearchDS}(f, \sigma_0, \epsilon_\sigma, y^*, \epsilon_y, \mathbf{x}, \epsilon_x, \hat{\mathbf{X}})$
6 $\quad$**until** *all optima found or max nb of function evaluations reached*

Algorithm 3: CMA-ES

{Covariance Matrix Adaptation Evolution Strategy}
Input:
$\quad f$: function to optimize
$\quad \lambda$: number of sample per iteration
$\quad x$: set of population
$\quad s$: set of fitness
Output: x_1: best optima find so far

1 **begin**
2 $\quad$Initialization **while** *stop criterion not met* **do**
3 $\quad\quad$**for** *i in 1...λ* **do**
4 $\quad\quad\quad x_i =$ sample
5 $\quad\quad\quad s_i = f(i)$
6 $\quad\quad\quad \text{Sort}(x_{1...\lambda})$
7 $\quad\quad\quad \text{Update\_mean\_to\_better\_solution}()$
8 $\quad\quad\quad \text{Update\_isotropic\_evolution\_path}()$
9 $\quad\quad\quad \text{Update\_anisotropic\_evolution\_path}()$
10 $\quad\quad\quad \text{Update\_covariance\_matrix}()$
11 $\quad\quad\quad \text{Update\_step-size}()$

Algorithm 4: CMA-ES-IPOP

{Covariance Matrix Adaptation Evolution Strategy with Increasing
$\quad\quad$POPulation } **Input:**
$\quad f$: function to optimize
$\quad \lambda$: number of sample per iteration
$\quad \lambda_0$: size of the initial population
$\quad n$: number of restart
$\quad X$: set of solution
Output: X or X_1 : set of optima or the best one

1 **begin**
2 $\quad n \leftarrow 0$ **while** *number of evaluations not reach* **do**
3 $\quad\quad n \leftarrow n + 1$
4 $\quad\quad \lambda \leftarrow \lambda_0 + 2 \times n$
5 $\quad\quad X \leftarrow \text{cmaes}(\lambda, f)$

All the results are reported in Table 1, and correspond to the best value found over the 100 runs, mean with confidence interval, and the median value. Both algorithms substantially improved the default settings of the experts. These results show the relevance of using data-oriented calibration with an ES algorithm on this kind of real-world application since it is able to find a set of parameters that allows the model to correctly predict sensor values. Moreover, according to the Mann-Whitney test at confidence level 0.01, the QRDS outperform the CMAES-IPOP on all crops. Figure 3 shows the dispersion of the values found overall runs (smaller is better). The restart strategy of QRDS seems to be more effective on such multi-model problem. Indeed, the exploration behavior of the QRDS allows finding more interesting search space area. As an example of the result, the Fig. 2 shows the predicted SWC by the best parameter settings of the Weedric model. In particular, the parameters setting improves the prediction for the dry period after the 30 days.

The second experiment consists in testing the robustness of an optimized parameter set. To do this, we select the parameter set that has obtained the best (smallest) fitness (one solution of the crop 4), then we use this set (from crop4) on the other crops. Table 2 presents the results of the corresponding fitness with the fitness of the best optimized solutions and default parameters. We can see that, indeed, results are not as good as a specific optimization,

Table 1. Best, median and mean fitness (with the confidence interval at 95%) found by each algorithm over 100 runs with a budget of 10^5 evaluations (smaller is better). The bolded median values are significantly better according to the Mann-Whitney at confidence level of 0.01.

Crop	Algorithm	Best	Mean	Median
Crop 1	Default	80.83	/	/
	QRDS	**15.8**	16.3 ± 0.03	**16.3**
	CMA-ES IPOP	16.4	18.5 ± 0.12	18.3
Crop 2	Default	57.81	/	/
	QRDS	**16**	16.7 ± 0.03	**16.7**
	CMA-ES IPOP	17	19 ± 0.08	19
Crop 3	Default	63.65	/	/
	QRDS	**17.5**	18 ± 0.03	**17.9**
	CMA-ES IPOP	19	21.5 ± 0.12	21.6
Crop 4	Default	47.85	/	/
	QRDS	**14.2**	15.8 ± 0.06	**16**
	CMA-ES IPOP	16.7	20 ± 0.25	19.4
Crop 5	Default	50.74	/	/
	QRDS	**15.1**	15.3 ± 0.02	**15.3**
	CMA-ES IPOP	15.3	16.6 ± 0.07	16.5

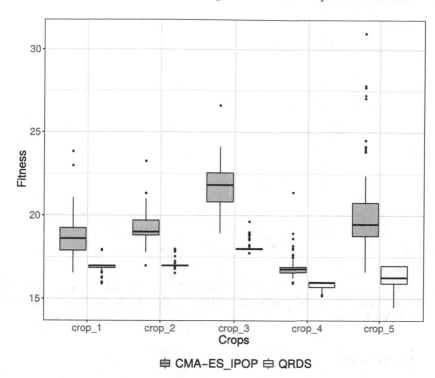

Fig. 3. Spread out of the best solutions found by QRDS and CMA-ES IPOP over 100 runs (smaller is better) on the 5 crops.

Table 2. Evaluation of the robustness of the best set of parameters found on crop 4 on the other crops compared to the fitness of their best optimized solution as well as the basic one (smaller is better). The best set of parameters ever found is not as good as the specific optimization of the problem but greatly improves fitness compared to the basic values.

Parameters	Crop1	Crop2	Crop3	Crop5
Default	80.83	57.81	47.85	50.74
Best Optim	15.8	16	17.5	15.1
Crop 4	38.27	25.78	31.8	17.71

but the solution is robust enough as it is by far better than the specific expert parameters. Figure 4 presents the results of the corresponding fitness with the fitness of the best optimized solutions and basic parameters. Without *a priori* knowledge, statistically, QRDS always finds better solutions than CMAES-IPOP.

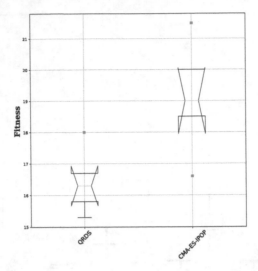

Fig. 4. Spread out of the mean fitness for QRDS (left) and CMA-ES IPOP (right) on the various crops. The average fitness of QRDS is statistically better (smaller) than that of CMA-ES IPOP.

5   Conclusion

In this paper, we propose to describe a farming decision-making model for irrigation into a black box optimization problem and we experiment on two state-of-the-art algorithms QRDS and CMA-ES IPOP. Results show that for this kind of problem using an ES algorithms is very efficient since independently of the crops, the set parameters calibrated by the algorithms are always significantly better than those by default (which have been designed by experts or available in the literature). Moreover, multi-modal QRDS seems to be very effective in such problems.

The proposed method is "offline" which means that the optimization of the parameter sets can only be done once the growing season is completed. Moreover, in agriculture, it is impossible to replant the same plant before 3 to 5 years on the same crop (and obviously the weather changes from year to year). This is for these reasons, that it is interesting to be able to find a robust solution, which may still be good for the next year. In order to have a solution as robust as possible, a possible future work could be to optimize different crops at the same time. Finally, we could also develop an "online" method able to predict the outputs of the model according to the already collected data and parameter sets.

Acknowledgements. The authors would like to thank the **WEENAT** company in particular for the financing of the CIFRE thesis and for their material support. Experiments presented in this paper were carried out using the CALCULCO computing platform, supported by SCOSI/ULCO (Service COmmun du Système d'Information de l'Université du Littoral Côte d'Opale).

References

1. Tang, Y., Reed, P., Wagener, T.: How effective and efficient are multiobjective evolutionary algorithms at hydrologic model calibration? Hydrol. Earth Syst. Sci. Discuss. **2**(6), 2465–2520 (2005)
2. Gupta, H.V., Beven, K.J., Wagener, T.: Model calibration and uncertainty estimation. In: Encyclopedia of Hydrological Sciences (2006)
3. Whisler, F.D., et al.: Crop simulation models in agronomic systems. In: Advances in Agronomy, vol. 40, pp. 141–208. Elsevier (1986)
4. Raes, D., Steduto, P., Hsiao, T.C., Fereres, E.: Aquacrop–the FAO crop model to simulate yield response to water: II. Main algorithms and software description. Agron. J. **101**(3), 438–447 (2009)
5. Ramat, E., Vandoorne, B.: Plant growth model for decision making support. Technical report, Université du Littoral Côte d'Opale, and ISA Lille (2002)
6. Beyer, H.-G.: The Theory of Evolution Strategies. Springer, New York (2001). https://doi.org/10.1007/978-3-662-04378-3
7. Rapin, J., Teytaud, O.: Nevergrad-a gradient-free optimization platform (2018). https://GitHub.com/FacebookResearch/Nevergrad
8. Rechenberg, I.: Evolutionsstrategie: Optimierung technischer Systeme nach Prinzipien der biologischen Evolution. Number 15 in Problemata. Frommann-Holzboog (1973)
9. Li, X.: Multimodal optimization using niching methods, pp. 1–8. American Cancer Society (2016)
10. Preuss, M.: Multimodal Optimization by Means of Evolutionary Algorithms. NCS. Springer, Cham (2015). https://doi.org/10.1007/978-3-319-07407-8_7
11. Ahrari, A., Deb, K., Preuss, M.: Multimodal optimization by covariance matrix self-adaptation evolution strategy with repelling subpopulations. Evol. Comput. **25**(3), 439–471 (2017)
12. Auger, A., Hansen, N.: A restart CMA evolution strategy with increasing population size. In: 2005 IEEE Congress on Evolutionary Computation, vol. 2, pp. 1769–1776. IEEE (2005)
13. Kadioglu, S., Sellmann, M., Wagner, M.: Learning a reactive restart strategy to improve stochastic search. In: Battiti, R., Kvasov, D.E., Sergeyev, Y.D. (eds.) LION 2017. LNCS, vol. 10556, pp. 109–123. Springer, Cham (2017). https://doi.org/10.1007/978-3-319-69404-7_8
14. Teytaud, F., Teytaud, O.: Qr mutations improve many evolution strategies: a lot on highly multimodal problems. In: Proceedings of the 2016 GECCO Conference, pp. 35–36 (2016)
15. Schoenauer, M., Teytaud, F., Teytaud, O.: A rigorous runtime analysis for quasi-random restarts and decreasing stepsize. In: Hao, J.-K., Legrand, P., Collet, P., Monmarché, N., Lutton, E., Schoenauer, M. (eds.) EA 2011. LNCS, vol. 7401, pp. 37–48. Springer, Heidelberg (2012). https://doi.org/10.1007/978-3-642-35533-2_4
16. Beaujouan, V.: Modélisation des transferts d'eau et d'azote dans les sols et les Nappes. Développement d'un modèle conceptuel distribué. Application à de petits bassins versants. Ph.D., thesis, Ecole Nationale Supérieure Agronomique de Rennes (2001)
17. Allen, R.G., Pereira, L.S., Raes, D., Smith, M., et al.: Crop Evapotranspiration-Guidelines for Computing Crop Water Requirements-FAO Irrigation and Drainage paper 56. FAO, Rome, vol. 300, no. 9 (1998). D05109

204   A. Dubois et al.

18. Teng, P.S., Johnson, K.B., Johnson, S.B.: Development of a simple potato growth model for use in crop-pest management. Agric. Syst. **19**(3), 189–209 (1986)
19. Beaujouan, V., Durand, P., Ruiz, L.: Modelling the effect of the spatial distribution of agricultural practices on nitrogen fluxes in rural catchments. Ecol. Model. **137**(1), 93–105 (2001)
20. Van Genuchten, M.T.: A closed-form equation for predicting the hydraulic conductivity of unsaturated soils 1. Soil Sci. Soc. Am. J. **44**(5), 892–898 (1980)
21. Hansen, N., Müller, S.D., Koumoutsakos, P.: Reducing the time complexity of the derandomized evolution strategy with covariance matrix adaptation (CMA-ES). Evol. Comput. **11**(1), 1–18 (2003)
22. Hansen, N., Kern, S.: Evaluating the CMA evolution strategy on multimodal test functions. In: Yao, X., et al. (eds.) PPSN 2004. LNCS, vol. 3242, pp. 282–291. Springer, Heidelberg (2004). https://doi.org/10.1007/978-3-540-30217-9_29

A Hybrid Evolutionary Algorithm for Offline UAV Path Planning

Soheila Ghambari[1(✉)], Lhassane Idoumghar[1(✉)], Laetitia Jourdan[2(✉)], and Julien Lepagnot[1(✉)]

[1] University of Haute-Alsace, IRIMAS Institute, Mulhouse, France
{soheila.ghambari,lhassane.idoumghar,julien.lepagnot}@uha.fr
[2] University of Lille, CRIStAL, UMR 9189, CNRS, Centrale Lille, Lille, France
laetitia.jourdan@univ-lille.fr

Abstract. This paper investigates the offline path planning problem of unmanned aerial vehicles (UAVs) for surveillance mission in complex urban environments. A new idea by coupling the differential evolution (DE) with A* algorithm is suggested to address the problem in large urban areas with narrow street and infrastructure of built environment. The proposed method consists of two phase: the first phase adopts DE to divide the straight line between source and destination into several smaller regions, while the second one utilizes A* for each region to find a collision-free and shortest path in parallel. In order to assess the efficiency of the suggested algorithm, a real-world scenario is examined. Evaluations exhibited promising results with proper accuracy and minimum computational time.

Keywords: UAV · Offline path planning · Differential evolution · A* algorithm

1 Introduction

The development of autonomous UAVs is of high interest to many military and civilian applications for various missions. In recent years, more studies focus on one of the essential aspect of UAV autonomy which is the capability for automatic path planning [7]. This process consists of finding an optimal or near-optimal collision-free path between the start and target positions; under specific constraints conditions. As a matter of fact, a suitable path planning strategy should be design not only to improve the effectiveness of the system (e.g., memory consumption and computational time) but also to communicate with other elements in order to comply with the mission requirements. Hence, implementing an effective algorithm entails a deep analysis of various contributing techniques [18].

Previous studies have presented a series of techniques to tackle the aforementioned problem based on different necessities such as performances optimization, collision avoidance, real-time planning, and safety maximization. They took

© Springer Nature Switzerland AG 2020
L. Idoumghar et al. (Eds.): EA 2019, LNCS 12052, pp. 205–218, 2020.
https://doi.org/10.1007/978-3-030-45715-0_16

hints from different research fields; like mathematics for graph-based and probabilistic approaches [2,15], physics for potential field algorithm [6], or computer science for artificial intelligence methods [5,17,22]. Generally, we can categorize the existing works into classical techniques (i.e., graph-based search methods, sampling-based approaches, potential field), computational intelligence (CI) methods, and hybrid approaches.

Graph-based searches (e.g., A* and Dijkstra) were developed to find the shortest path between two nodes of connected graphs with a greedy logic. One of the positive characteristic of these methods is their simplicity, which implies reduced computational time. They have deterministic nature and guarantee to find the optimal collision-free path, if it exists. However, the performance of these algorithms depends on the environment's total area due to the fact that they save all explored nodes in memory. Sampling-based methods, such as Probabilistic Roadmaps (PRM) [13] and Rapidly-exploring Random (RRT) [14] have proven to be an effective framework suitable for high-dimensional spaces to produce feasible solutions; nevertheless, they do not guarantee the optimality of the solution [9]. In recent years, CI methods including fuzzy system, neural network, and evolutionary algorithms (EAs) have received most of the research effort for solving UAV path planning problem [10,21]. They attract the attention of researchers because of: (a) their flexibility to solve large-scale complex problems, (b) their ability to apply different learning strategies to perform an effective search towards the global optimum, and (c) employing for both single and multiple UAVs. However, in practice, when the available computation resources and/or time are limited, they are not always the best choice.

These issues motivated us to present a novel hybrid approach inspired from incremental heuristic search which not only scales well with problem size but also speeds up the search process for a high quality path in a reasonable execution time. To do so, A* as an informed heuristic search strategy and DE algorithm as one of the most popular EAs are integrated in order to find the shortest collision-free path in high dimension spaces with minimum computational time. In this method, the search space is limited around the straight line between the start and target locations. This is motivated by the fact that taking into account the whole configuration space can raise the computational cost. Here, the start and the target points are connected to each other via a straight line (as the shortest path) regardless of the obstacles. Then, we apply DE algorithm to divide the straight line into several suitable segments/regions. Thereafter, A* is used as a local path planner to find the shortest path for each region in a parallel manner. Altogether, the suggested method reduce the dimensionality of the search space which enables the presented algorithm to find better topologically distinct paths more rapidly. The performance of the proposed method is compared with A* and standard DE algorithm for a realistic urban environment. Evaluations exhibited desirable run-time performance in finding feasible and safe paths.

The rest of this paper is organized as follows. Section 2 starts with problem definition in Subsect. 2.1. Next, in Subsects. 2.2 and 2.3 basic concepts of A* and DE algorithm are explained, respectively. The description of the proposed method including environment modeling, constraints, solution representation, objective function, and the suggested algorithm are provided in Sect. 3. The simulation results and discussion are presented in Sect. 4. Finally, the paper is concluded in Sect. 5.

2 Background Information

2.1 Problem Definition

Generally, path planning belongs to a class of non-deterministic polynomial-time (NP) hard problems [4] which is much more intensively investigated in robotics (referred as motion planning). Formally, path planning for UAVs defined as an optimization problem aimed at finding the shortest and safest path to reach a goal position, while flying into a high-threat area. Here, some important factors should be taken into consideration such as modeling the environment, the path representation, safety, cost of the path, and computational time. These factors are either integrated directly into the objective functions that require to be minimized/maximized, or in the form of constraints that a path must comply with. The later subsections elaborates these factors with more details.

2.2 A* Search Algorithm

The history of finding the shortest path can be followed as early as 1968, when A* as the most effective direct search method is developed for robot navigation [8,12]. The algorithm acts on the basis of *Dijkstra*, but can avoid blind search to improve search efficiency. It seeks towards the most promising states using a heuristic function in order to save the computational time resource. A detailed explanation of A* can be found in [12].

2.3 Differential Evolution

DE algorithm has been successfully employed in various research and application areas. It has been also utilized in path planning tasks for both single UAV and multiple UAVs [3,19,20]. DE is an iterative procedure which aims at evolving a population (NP) of D-dimensional parameter vectors towards the global optimum. It includes a population of path candidate solutions or individuals which are produced by integrating a parent and other individuals of the same population. Each candidate solution has a set of variables which subjected to mutation and crossover search operators in order to produce new solutions subject to some constraints. The algorithm only accepts the candidate solutions that are better than their parents and accordingly transfers them to the next generation of the algorithm. The algorithmic description is summarized in Fig. 1. In this figure, the five most frequently utilized mutation strategies are listed.

The indices r_1, r_2, r_3, r_4, r_5 are mutually exclusive integers randomly generated within the range $[1, NP]$, which also differ from the index i. These indices

The DE Algorithm Pseudo Code

Generate the initial population
Evaluate the fitness for each individual
While the stopping criterion is not satisfied do
For i =1 to NP **do**
  Select three mutually different individuals $r_1 \neq r_2 \neq r_3$
    $\neq i$
  $j_{rand} = int\,(rand\,[1, D])$
  For j =1 to D **do**
    If $rand\,[0, 1] < CR$ or $j = j_{rand}$ **then**
    $u_{i,G}^j$ Apply the predetermined strategy
    Else
      $u_{i,G}^j = x_{i,G}^j$
    End if
  End for

End for
For i =1 to NP **do**

  Evaluate the offspring
  If the fitness function value of $u_{i,G}$ is no worse than
    $x_{i,G}$ **then** replace $x_{i,G}$ with $u_{i,G}$
  End if
End for
End while

Different type of strategy applied in DE algorithm
DE/rand/1
$u_{i,G} = x_{r1,G} + F\,(x_{r2,G} - x_{r3,G})$
DE/best/1
$u_{i,G} = x_{best,G} + F\,(x_{r1,G} - x_{r2,G})$
DE/rand-to-best/1
$u_{i,G} = x_{i,G} + F\,(x_{best,G} - x_{i,G})$ $+ +F\,(x_{r1,G} - x_{r2,G})$
DE/best/2
$u_{i,G} = x_{best,G} + F\,(x_{r1,G} - x_{r2,G})$ $+ +F\,(x_{r3,G} - x_{r4,G})$
DE/rand/2
$u_{i,G} = x_{r1,G} + F\,(x_{r2,G} - x_{r3,G})$ $+ +F\,(x_{r4,G} - x_{r5,G})$

Fig. 1. The pseudo code of DE algorithm

are randomly generated once for each mutant vector. The scaling factor F is a positive control parameter for scaling the difference vector. The crossover rate CR is a user-specified fixed within the range $[0, 1)$, which controls the fraction of parameter values copied from the mutant vector. $X_{best,G}$ is the best individual vector with the best fitness value in the population at generation G. j_{rand} is a randomly chosen integer in the range $[1, D]$.

3   The Proposed Approach

First, a clear description of environment modeling, constraints, solution representation, and objective function is presented. Thereafter, the introduced algorithm is described in details.

3.1   Environment Modeling and Constraint

In this work, we considered a grid-based map to represent the environment. The grid map is composed of equal size cells, where each cell is represented by a unique number. An urban environment in a 2-dimensional (2D) form is pre-processed to generate the grid map. In this step, an occupancy matrix is utilized for grid map representation where each cell has two possible values: "0" for a free and "1" for an occupied cell. The buildings with different polygon shapes are considered as obstacles; which are static and known in advance. In order to understand how these polygon shapes occupy the grid cells, polygon triangulation method is used to decompose a polygon area into a set of triangles with pairwise non intersecting interiors. Accordingly, we check whether a grid cell lies inside a triangle or not (see Fig. 2). The occupancy matrix is pre-processed only once and the back-tracking process for making paths does not consume significant computational resources.

The constraint is path safety which means that a path always should satisfy a predefined safety margin (distance) with respect to the obstacles. In this work, the safety margin is the confidence radius of UAV around obstacles which is considered as 1 unit.

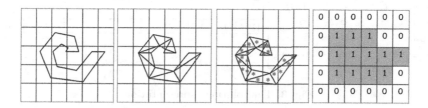

Fig. 2. The obstacle modeling in a grid map representation (The first three figures, on the *left* side, show how an arbitrary polygon shape occupies grid cells using the triangulation method. The occupancy map is represented in the *right* side figure.)

3.2 Solution Representation

The solution representation is an essential element for solving an optimization problem. In this study, each solution/path consists of a sequence of design variables. These variables are adopted based on grid cells that are located in the straight line between the start and target positions; with their unique numbers. In this way, the algorithm focuses on the most promising parts of the search space which can enhance the convergence performance. If a variable did not satisfy the constraint, the perpendicular line that passes through the selected variable is considered and another arbitrary point upon this line which is collision-free and near to the straight line will be chosen. An example of modeling the configuration space and solution representation is displayed in Fig. 3.

3.3 Objective Function

The objective function has to satisfy the constraints while optimizing the flight path and avoiding obstacles. Here, owing to employing grid map representation a feasible flight path can be defined from the start to the destination cell by traversing a certain number of free cells [1]. Hence, the cost of a feasible path is the sum of all costs of the movements along the associated path in all regions. The UAV is assumed to move horizontally or vertically or diagonally from a free cell to another one with fixed flight altitude. Accordingly, there are eight

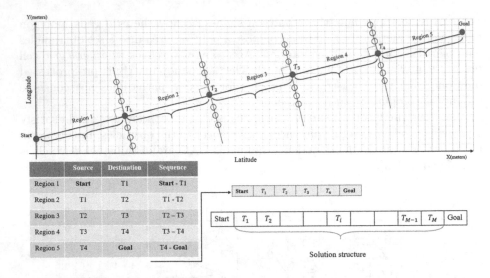

Fig. 3. The solution representation

possible moves from each cell to another one. It is worth mentioning that the main contribution of this work is to apply A* heuristic objective function to find the shortest path length in a desirable run-time.

3.4   ADE Algorithm

As mentioned before, adopting a fast and efficient path planning method is critical for autonomous UAVs which usually operate in large scale urban environments. There are various intelligent optimization methods which have been successfully applied in solving UAV path planning problem [11,21]. In the same direction and without loss of generality, this paper presents a new approach by integrating A* and DE algorithm in order to generate the shortest path with minimum computational time over very long distances.

The introduced hybrid algorithm, named as ADE, contains two main steps. The first step is accomplished with DE which is responsible for intersecting the whole area into conjunct regions. In fact, it determines several intermediate cells for exploring better the configuration space. These cells are DE's design variables which are located on the straight line between the start and target positions. In this way, the algorithm focuses on the most promising parts of the search space which can enhance its convergence performance. As explained in previous subsection, if a variable did not satisfy the constraint, a straight line perpendicular to the selected variable is considered, and another arbitrary point upon this line will be chosen by DE. This point should have two conditions: (a) be in an admissible space, (b) have a minimum distance from the straight line. In such a way, a proper balance between the exploration and exploitation capabilities of DE algorithm is achieved. In the second step, A* algorithm is employed to find the shortest path in each region in parallel. Thus, all the paths obtained from regions are connected to form the global best path. Hence, the algorithm is widely favorable for reducing computational time. An example of modeling the configuration space and the proposed algorithm is displayed in Fig. 4. Moreover, the flowchart of the proposed algorithm is shown in Fig. 5.

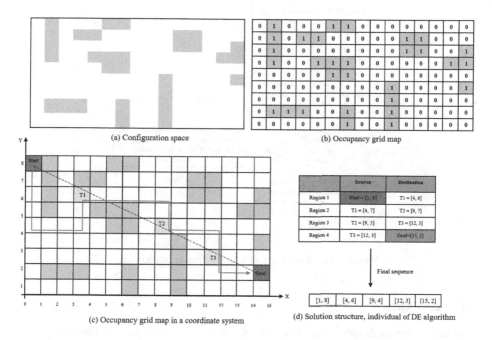

(a) Configuration space   (b) Occupancy grid map

(c) Occupancy grid map in a coordinate system   (d) Solution structure, individual of DE algorithm

Fig. 4. An example of the proposed path planning algorithm: (a) configuration space, (b) occupancy grid map, (c) an obtained path in a coordinate system, (d) solution representation

4   Experimental Evaluation

This section aims to investigate the efficiency of the presented algorithm through a series of experiments on a realistic urban environment. The selected test case provided the chance to conduct a comprehensive study on the performance of algorithm in terms of path length and computational time. For this purpose, Subsect. 4.1 begins with a description of the test case characteristics. Then, in Subsect. 4.2, the setting parameters are introduced. In order to automatically configure the algorithm's parameters, *irace* package is utilized. Finally, the compared algorithms and statistical results obtained via experiments on urban map are presented and discussed in Subsects. 4.3 and 4.4, respectively. All simulations and evaluations were implemented and conducted within Python library[1], on a computer with Intel Core i5-7440HQ CPU, 2.80 GHz, 8GB RAM, running on Ubuntu OS.

---

[1] (Atsushi Sakai et al. https://github.com/AtsushiSakai/PythonRobotics).

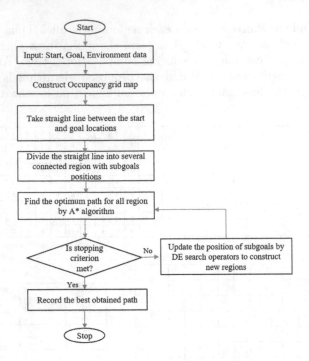

Fig. 5. The flowchart of the proposed algorithm

4.1   Test Case

The experiments have been extended with realistic urban environment. The selected environment for evaluating the performance of the algorithm is a partial part of Mulhouse city in France. The map file is extracted from OpenStreetMap, defined by geographical coordinates in terms of latitude and longitude. In this file the buildings tags are filtered. These building are taken into account as obstacles and their modeling is explained in Subsect. 3.1. The characteristics of this map are summarized in Table 1. In addition, Fig. 6 shows the total map and part of its modeling. As can be seen from part of modeling, this map has narrow streets with compressed obstacles.

Table 1. The test case characteristics

Map	Latitudes	Longitudes	No. obstacles
Mulhouse	Minlat = 47.7250	Minlon = 7.3001	4099
	Maxlat = 47.7538	Maxlon = 7.3466	

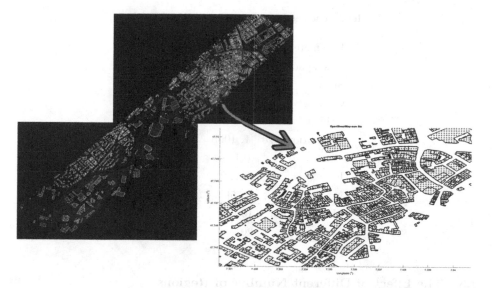

Fig. 6. Illustration of the selected map and employed polygon triangulation method

4.2   Experimental Setup

The configuration parameters of the algorithm can be divided into two categories: environment and algorithm parameters. Environment parameters include: grid size, start point, target point, number of obstacles and their coordination, and the boundary of the search space which are the minimum and maximum of latitudes and longitudes. The algorithm parameters are population size, crossover probability (CR), scaling factor (F), type of DE strategies, number of design variables, and number of iterations. The number of variables, which is assumed as dimension of the problem or regions, is an integer within the range $[1, 9]$. If the algorithm adopts 1, it means A* algorithm is applied for the total configuration space. Also, the maximum number of iterations and runs for this work are 100 and 20, respectively.

Table 2 describes the configurable settings of the proposed algorithm. As mentioned above, some parameter settings including population size, CR, F, and type of DE strategy are significant for a certain value and have a great impact on the performance of algorithm. Hence, instead of using a trial-and-error approach to identify good values for these parameters, we utilized *irace* software [16] as an automated algorithm configuration tool for obtaining very high-performing algorithmic variants. A maximum budget of 200 experiments is applied for each run of *irace* and it is repeated 20 times to assess the variability of the automatic configuration process. According to the obtained results, the best configuration uses $[80, 1, 0.2, 1]$ values for population size, F, CR, and the type of strategy parameters, respectively. The related DE mutation strategy, labelled by the number 1 during the parameter setting, is DE/rand/1.

Table 2. The setting parameter of the algorithm

Parameter	Value
Grid space	461 * 286
Start	[x = 84, y = 40]
Goal	[x = 387, y = 251]
Grid size	1
Population size	[1, 100]
CR	[0.1, 1]
F	[0.1, 2]
No. strategy	[1, 5]
Max iteration	100
Max run	20

4.3   The Effect of Different Number of Regions

As was mentioned before, the presented approach divides the distance between the start and target locations into several regions. The number of these regions which are taken into account as the number of dimension are very important to be determined. Thus, to investigate whether this number has a positive effect on the performance of the algorithm in terms of precision of path length and computational time, a comparison using different number of regions is performed. Parameter configurations for this experiment are similar to the settings explained in the previous subsection. Figure 7 exhibits the average and standard deviation of path length and computational time for different number of regions over 20 independent runs. As can be seen, by increasing the number of dimension the computational time significantly decreased; while as expected the precision of path length is approximately reduced. Furthermore, standard deviation of the results shows the stability of the presented method. As a matter of fact, it clearly confirms that the difference between regions can considerably affect the computational time which is very important factor especially in large scale environment. Another interesting observation that can be concluded from these results is that this approach makes the problem as a low dimensional problem using less number of decision variables for dividing the regions.

4.4   Results and Discussion

The presented algorithm was executed over 20 independent runs. The results are presented by the best, mean, and standard deviation (S.D.) of cost values obtained in all runs. To provide a meaningful comparison of A*, DE, and ADE, the mean and S.D. of the path length and computational time are compared with each other. All experimental results are reported in Table 3.

As it was expected, the results of ADE shows the impact of adopting different number of regions in accuracy of path length and computational time. ADE

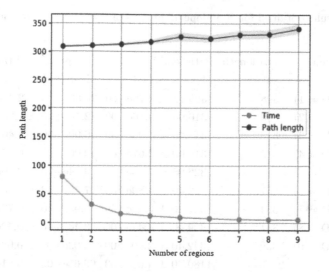

Fig. 7. The effect of different values for number of regions

with two regions has a smaller path length than the other dimension. However, its computational time is greater. The other dimensions have a close competition in accuracy of path length where by increasing the number of regions, the computational time surprisingly reduced. Also, the result of original DE shows that this algorithm was not able to find the shortest path in a reasonable time. One of the reason for such bad performance is the small number of iterations that makes it hard for DE to find the best set of grid cells. Finally, the results of

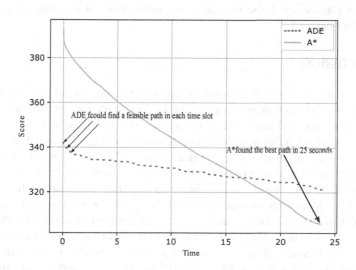

Fig. 8. The obtained feasible path in a predetermined time slot for both A* and ADE algorithm

Table 3. Results obtained for 20 independent runs of algorithms for offline path planning

Algorithm	Path length	Path length	Computational time (s)
	Best	Mean ± S.D.	Mean ± S.D.
ADE (D = 1)	308	309.12 ± 0.76e+00	38.60e+00 ± 2.43e+00
ADE (D = 2)	309	310.05 ± 0.76e+00	32.10e+00 ± 1.25e+00
ADE (D = 3)	310	313.40 ± 1.95e+00	16.00e+00 ± 1.40e+00
ADE (D = 4)	312	316.00 ± 2.67e+00	12.00e+00 ± 0.72e+00
ADE (D = 5)	315	326.05 ± 7.56e+00	9.14e+00 ± 0.80e+00
ADE (D = 6)	317	322.40 ± 5.80e+00	7.70e+00 ± 0.65e+00
ADE (D = 7)	324	327.40 ± 6.51e+00	3.98e+00 ± 0.73e+00
ADE (D = 8)	331	329.40 ± 4.42e+00	3.46e+00 ± 0.52e+00
ADE (D = 9)	339	342.40 ± 2.60e+00	3.03e+00 ± 0.08e+00
DE	2170	1180.00 ± 4.09e+03	66.65e+00 ± 3.81e+00
A*	307	307.00 ± 0.00e+00	13.80e+00 ± 0.16e+00

A* algorithm is reported in the last row of this table. A* could find the shortest path with high accuracy but with more computational time when compared to the presented ADE algorithm. There is a close competition between A* and ADE with dimension 4.

Also, Fig. 8 shows that the proposed method can give more accurate solutions in the early iterations, while A* finds the best possible flyable path using more computational resource. Indeed, this is the main properties of ADE which allows us to make a trade-off between these two conflicting objectives; as previously explained in Sect. 4.3.

5   Conclusion

This study concerns the development of a new path planning algorithm for UAVs so as to avoid obstacles in realistic urban environment for surveillance mission. The problem is modeled in a static 2D space with constraint single objective function. The suggested ADE approach integrated a heuristic search function with DE for large-scale environments. For this purpose, DE is employed to divide the configuration space into several conjunct regions. Then, each region is explored by A* algorithm as a local path planner in a parallel manner. The presented algorithm tries to search around the straight line between the start and destination which results in increasing the convergence performance and decreasing the computational time. The performance of the algorithm is evaluated through a series of experiments. The *irace* software package is also utilized in order to find the best configurations for the algorithm. The obtained results illustrated the efficiency of ADE in finding optimal solutions with proper accuracy and minimum computational time.

Acknowledgment. This work is part of a project funded by the French *Agence Nationale de la Recherche* under grant number ANR-16-SEBM-0004.

References

1. Alajlan, M., Koubâa, A., Châari, I., Bennaceur, H., Ammar, A.: Global path planning for mobile robots in large-scale grid environments using genetic algorithms. In: 2013 International Conference on Individual and Collective Behaviors in Robotics (ICBR), pp. 1–8. IEEE (2013)
2. Bauso, D., Giarré, L., Pesenti, R.: Multiple UAV cooperative path planning via neuro-dynamic programming. In: 2004 43rd IEEE Conference on Decision and Control (CDC) (IEEE Cat. No. 04CH37601), vol. 1, pp. 1087–1092. IEEE (2004)
3. Brintaki, A.N., Nikolos, I.K.: Coordinated UAV path planning using differential evolution. Oper. Res. **5**(3), 487–502 (2005)
4. Canny, J.: The Complexity of Robot Motion Planning. MIT Press, Cambridge (1988)
5. Cekmez, U., Ozsiginan, M., Sahingoz, O.K.: A UAV path planning with parallel ACO algorithm on CUDA platform. In: 2014 International Conference on Unmanned Aircraft Systems (ICUAS), pp. 347–354. IEEE (2014)
6. Chen, Y., Luo, G., Mei, Y., Yu, J., Su, X.: UAV path planning using artificial potential field method updated by optimal control theory. In. J. Syst. Sci. **47**(6), 1407–1420 (2016)
7. Choi, Y., Choi, Y., Briceno, S., Mavris, D.N.: Energy-constrained multi-UAV coverage path planning for an aerial imagery mission using column generation. J. Intell. Robot. Syst. **97**(1), 125–139 (2019). https://doi.org/10.1007/s10846-019-01010-4
8. Dijkstra, E.W.: A note on two problems in connexion with graphs. Numerische mathematik **1**(1), 269–271 (1959)
9. Gammell, J.D., Srinivasa, S.S., Barfoot, T.D.: Informed RRT*: optimal sampling-based path planning focused via direct sampling of an admissible ellipsoidal heuristic. In: 2014 IEEE/RSJ International Conference on Intelligent Robots and Systems, pp. 2997–3004. IEEE (2014)
10. Ghambari, S., Lepagnot, J., Jourdan, L., Idoumghar, L.: A comparative study of meta-heuristic algorithms for solving UAV path planning. In: 2018 IEEE Symposium Series on Computational Intelligence (SSCI), pp. 174–181, November 2018. https://doi.org/10.1109/SSCI.2018.8628807
11. Goerzen, C., Kong, Z., Mettler, B.: A survey of motion planning algorithms from the perspective of autonomous UAV guidance. J. Intell. Robot. Syst. **57**(1–4), 65 (2010)
12. Hart, P.E., Nilsson, N.J., Raphael, B.: A formal basis for the heuristic determination of minimum cost paths. IEEE Trans. Syst. Sci. Cybern. **4**(2), 100–107 (1968)
13. Kavraki, L., Svestka, P., Overmars, M.H.: Probabilistic roadmaps for path planning in high-dimensional configuration spaces. IEEE Trans. Robot. Autom. **12**, 566–580 (1994)
14. LaValle, S.M., Kuffner Jr., J.J.: Randomized kinodynamic planning. Int. J. Robot. Res. **20**(5), 378–400 (2001)
15. Li, J., Sun, X.: A route planning's method for unmanned aerial vehicles based on improved a-star algorithm. Acta Armamentarii **7**, 788–792 (2008)
16. López-Ibáñez, M., Dubois-Lacoste, J., Stützle, T., Birattari, M.: The irace package, iterated race for automatic algorithm configuration. Technical report, TR/IRIDIA/2011-004, IRIDIA, Université Libre de Bruxelles (2011)

17. Ji, X.-T., Xie, H.-B., Zhou, L., Jia, S.-D.: Flight path planning based on an improved genetic algorithm. In: 2013 Third International Conference on Intelligent System Design and Engineering Applications, pp. 775–778. IEEE (2013)
18. Yang, P., Tang, K., Lozano, J.A., Cao, X.: Path planning for single unmanned aerial vehicle by separately evolving waypoints. IEEE Trans. Robot. **31**(5), 1130–1146 (2015)
19. Zhang, X., Duan, H.: An improved constrained differential evolution algorithm for unmanned aerial vehicle global route planning. Appl. Soft Comput. **26**, 270–284 (2015)
20. Zhang, X., Chen, J., Xin, B., Fang, H.: Online path planning for UAV using an improved differential evolution algorithm. IFAC Proc. Volumes **44**(1), 6349–6354 (2011)
21. Zhao, Y., Zheng, Z., Liu, Y.: Survey on computational-intelligence-based UAV path planning. Knowl.-Based Syst. **158**, 54–64 (2018)
22. Zhu, Y., et al.: Target-driven visual navigation in indoor scenes using deep reinforcement learning. In: 2017 IEEE International Conference on Robotics and Automation (ICRA), pp. 3357–3364. IEEE (2017)

Author Index

Printed in the United States
By Bookmasters